Windows 11 for Enterprise Administrators

Unleash the power of Windows 11 with effective techniques and strategies

Manuel Singer

Jeff Stokes

Steve Miles

Thomas Lee

Richard Diver

BIRMINGHAM—MUMBAI

Windows 11 for Enterprise Administrators

Group Product Manager: Pavan Ramchandani

Publishing Product Manager: Prachi Sawant

Book Project Manager: Neil Dmello

Senior Editor: Arun Nadar and Athikho Sapuni Rishana

Technical Editor: Irfa Ansari

Copy Editor: Safis Editing

Language Support Editor: Ashwin Kharwa

Proofreader: Safis Editing

Indexer: Hemangini Bari

Production Designer: Prafulla Nikalje

Marketing Coordinators: Marylou De Mello and Shruthi Shetty

First published: September 2017

Second edition: September 2023

Production reference: 1131023

Published by Packt Publishing Ltd.
Grosvenor House
11 St Paul's Square
Birmingham
B3 1RB

ISBN 978-1-80461-859-2

www.packtpub.com

Contributors

About the authors

Manuel Singer works as a Surface cloud solution architect at Microsoft and is based in Germany. He has over 20 years of experience in system management and deployment using Microsoft technologies and has worked for more than a decade for Microsoft. He helps Surface and Surface Hub customers to get the best experience with these devices and small, medium, and large-sized organizations transition from Windows 10 to Windows 11, AD to AAD, and Intune to Autopilot to get the most out of Microsoft 365. He specializes in client enterprise design, deployment, performance, reliability, and Microsoft devices. Manuel works with local and international top customers from the private and public sectors to provide professional technical and technological support.

First and foremost, I would like to dedicate this book to my family, especially to my wife, Renate, for her patience and continued support in allowing me the time to research and write this book. She is the reason I can fulfill my dream and follow my passion. I would also like to extend an acknowledgment to all the people who have supported me throughout the writing of this book, especially the technical reviewers, for providing their efforts and time along with keen suggestions and recommendations. Last but not least, I would like to thank the entire Packt Publishing team for their support and guidance throughout the process of writing this book.

Jeff Stokes works at Tanium as a Distinguished Engineer. He's an author, blogger, gamer, dad, and husband. He has worked in most areas of IT operations in the last 29 years, specializing in Windows debugging and performance analysis. His side interests include SANs, NLP, VDI configuration and optimization, and Windows imaging and deployment.

I want to thank the people in my life who have supported and encouraged my career over the years, including Vince Zolkosky, Carl Luberti, Clint Huffman, Yong Rhee, Mark Rowe, my family and friends, and my wife, Ana.

Steve Miles is a Microsoft security and Azure/hybrid MVP and MCT with over 20 years of experience in security, networking, storage, end user computing, and cloud solutions. His current focus is on securing, protecting, and managing identities, Windows clients, and Windows server workloads in hybrid and multi-cloud platform environments. His first Microsoft certification was on Windows NT and he is an MCP, MCITP, MCSA, and MCSE for Windows and many other Microsoft products. He also holds multiple Microsoft Fundamentals, Associate, Expert, and Specialty certifications in Azure security, identity, network, M365, and D365. He also holds multiple security, networking vendor, and other public cloud provider certifications.

This book is my contribution to the worldwide technical learning community, and I would like to thank all of you who are investing your valuable time in learning new skills and committing to reading this book.

Thomas Lee is a consultant, trainer, and writer from England. He has been in the IT world since the late 1960s and has worked around the world. Thomas was one of the first to discover PowerShell and has written numerous books on the subject. He has worked for a range of small as well as global firms, including Microsoft. He is semi-retired and spends his time with his family.

I love PowerShell and enjoy helping others to see the value of this amazing technology. If you are new to PowerShell, I hope you get to grips with managing Windows 11 using PowerShell.

Richard Diver is a senior technical business strategy manager for the Microsoft Security Solutions group, focused on developing security partners. Based in Chicago, Richard works with advanced security and compliance partners to help them build solutions across the entire Microsoft platform, including Microsoft Sentinel, Microsoft Defender, Microsoft 365 security solutions, and many more. Prior to Microsoft, Richard worked in multiple industries and for several Microsoft partners to architect and implement cloud security solutions for a wide variety of customers around the world. Any spare time he gets is usually spent with his family.

About the reviewer

Anton Romanyuk is an accomplished IT professional with years of experience in enterprise IT. He specializes in Windows devices, virtualization, and automation. Currently, Anton holds the position of cloud solutions architect at Microsoft, where he ensures customers have a great experience with Windows devices at work. Beyond his work at Microsoft, Anton is an enthusiastic and committed media informatics graduate with a strong background in freelance design. He also has a notable history in the indie gaming industry, having co-founded the highly successful *Wing Commander Saga: The Darkest Dawn* project in 2002. Anton's dedication to the project was unwavering, and his constant advocacy was instrumental in bringing the project to fruition.

The act of reviewing this book is Anton's sincere gesture of "giving back" to the IT community, which has imparted a number of techniques to him. Anton considers it his responsibility to share his knowledge with others, as he firmly believes that sharing is the key to community growth. He takes great pleasure in sharing his experiences in such a creative, engaging, and innovative environment as enterprise IT.

Table of Contents

2

Introduction to PowerShell 51

3

Configuration and Customization 79

4

User Account Administration   95

5

Tools to Manage Windows 11   109

6

Device Management   135

10

Windows 11 21H2 and 22H2 Changes (versus Windows 10)   237

Preface

Microsoft's launch of Windows 11 is a step toward satisfying Enterprise administrator needs for management and user experience customization. With its improvements and continuous developments since the first Windows 10 releases, it represents the latest and most secure Windows version to date. But to make the most of this security, it is necessary to deal with the new requirements and changes of Windows 11. This book provides Enterprise administrators with the knowledge required to fully utilize the advanced feature set of Windows 11 Enterprise. This practical guide shows Windows 11 from an administrator's point of view.

Who this book is for

If you're a system administrator tasked with upgrading to Windows 11, then this book is for you. Having deployed and managed previous versions of Windows in the past will help you follow along with this book, but you can also use it as a guide if Windows 11 is your first foray into system administration.

What this book covers

Chapter 1, *Windows 11 – Installation and Upgrading*, covers concepts and best practices for installing the new Windows 11 to prepare you for the move to Windows 11 in the most feasible way. It covers the new hardware requirements for Windows 11 and discusses different installation options. The chapter shows under which conditions an in-place upgrade is possible. It will also explain the new Lifecycle Policy of Windows 11.

Chapter 2, *Introducing PowerShell*, provides an introduction to PowerShell/PowerShell 7. The chapter explains the key concepts and shows how you can learn more about PowerShell.

Chapter 3, *Configuration and Customization*, discusses configuring Windows 11 to your needs, supported customization options, and how to configure Windows 11 for end users.

Chapter 4, *User Account Administration*, covers the administration of user accounts in Windows 11, including Azure AD, local accounts, and domain accounts usage.

Chapter 5, *Tools to Manage Windows 11*, discusses two sets of tools that you can use to manage Windows 11 and your Windows Server environment. **Remote Server Admin Tools** (**RSAT**) are produced by Microsoft and are available to download and use. The chapter also discusses the Sysinternals tools from Microsoft. Both tool sets are invaluable – and are both free and easy to obtain.

Chapter 6, *Device Management*, describes the new **mobile device management** (**MDM**) capabilities of Windows 10 and 11, discusses caveats of the Windows 10/11 GPO processing, and has a deeper look at patching and servicing, including the deployment solutions of the needed quality and feature updates such as Windows Update for Business, WSUS, MECM (aka SCCM), and third-party solutions.

Chapter 7, *Accessing Enterprise Data in BYOD and CYOD Scenarios*, covers an understanding of **Bring Your Own Device** (**BYOD**) and **Choose Your Own Device** (**CYOD**) models. You will see and understand how to handle the scenario of user access to corporate data on personally owned Windows 11 devices.

Chapter 8, *Windows 11 Security*, covers all aspects of Windows 11 security. While you have covered some aspects of security in some of the other chapters in this book, you will look at them collectively and in more detail in this security-focused chapter. If you are a security professional, then this chapter is dedicated to your role and responsibilities in securing Windows 11 in a company.

Chapter 9, *Advanced Configurations*, goes over a variety of different configurations used in enterprise environments, including VDI, kiosk mode, Autopilot, configuration for schools, Unbranded Boot, and WSL2.

Chapter 10, *Windows 11 21H2 and 22H2 Changes (versus Windows 10)*, gives an overview of all the new features and the numerous changes of the first two Windows 11 versions compared to the previous Windows 10 versions. This chapter is intended to give an overview of all the new features you should take a look at. It is a good start to familiarize yourself with the new features.

To get the most out of this book

We recommend that you install and activate a copy of Windows 11 Enterprise in a test environment. An Active Directory domain is required in order to test new Group Policy options. An Azure subscription is required to test the following features covered in this book:

- Azure Active Directory domain join
- Microsoft Intune for device management and Autopilot
- Security Center for Microsoft Endpoint Protection (MDE)

You may also want a Microsoft 365 E5 trial subscription to see the full potential and complete integration.

Conventions used

There are a number of text conventions used throughout this book.

`Code in text`: Indicates code words in text, database table names, folder names, filenames, file extensions, pathnames, dummy URLs, user input, and Twitter handles. Here is an example: "You can check SLAT availability via `coreinfo.exe -v` from Sysinternals Suite."

A block of code is set as follows:

```
Install-Script -Name Get-WindowsAutoPilotInfo -Force
Get-WindowsAutoPilotInfo -OutputFile AutoPilotInfo.csv
```

When we wish to draw your attention to a particular part of a code block, the relevant lines or items are set in bold:

```
PowerShell.exe -ExecutionPolicy Bypass
Install-Script -name Get-WindowsAutopilotInfo -Force
Set-ExecutionPolicy -Scope Process -ExecutionPolicy RemoteSigned
Get-WindowsAutoPilotInfo -Online
```

Bold: Indicates a new term, an important word, or words that you see onscreen. For instance, words in menus or dialog boxes appear in **bold**. Here is an example: "From Windows 11, you can use the Microsoft Store and search for PowerShell 7."

> **Tips or important notes**
> Appear like this.

Get in touch

Feedback from our readers is always welcome.

General feedback: If you have questions about any aspect of this book, email us at `customercare@packtpub.com` and mention the book title in the subject of your message.

Errata: Although we have taken every care to ensure the accuracy of our content, mistakes do happen. If you have found a mistake in this book, we would be grateful if you would report this to us. Please visit `www.packtpub.com/support/errata` and fill in the form.

Piracy: If you come across any illegal copies of our works in any form on the internet, we would be grateful if you would provide us with the location address or website name. Please contact us at `copyright@packt.com` with a link to the material.

If you are interested in becoming an author: If there is a topic that you have expertise in and you are interested in either writing or contributing to a book, please visit `authors.packtpub.com`.

Share Your Thoughts

Once you've read *Windows 11 for Enterprise Administrators*, we'd love to hear your thoughts! Scan the QR code below to go straight to the Amazon review page for this book and share your feedback.

https://packt.link/r/1804618594

Your review is important to us and the tech community and will help us make sure we're delivering excellent quality content.

Download a free PDF copy of this book

Thanks for purchasing this book!

Do you like to read on the go but are unable to carry your print books everywhere? Is your eBook purchase not compatible with the device of your choice?

Don't worry, now with every Packt book you get a DRM-free PDF version of that book at no cost.

Read anywhere, any place, on any device. Search, copy, and paste code from your favorite technical books directly into your application.

The perks don't stop there, you can get exclusive access to discounts, newsletters, and great free content in your inbox daily

Follow these simple steps to get the benefits:

1. Scan the QR code or visit the link below

https://packt.link/free-ebook/9781804618592

2. Submit your proof of purchase

3. That's it! We'll send your free PDF and other benefits to your email directly

1

Windows 11 – Installation and Upgrading

In this chapter, you will learn the concepts and best practices for installing the new **Windows 11** to prepare you for the move to Windows 11 in the most feasible way. We will cover the new hardware requirements for Windows 11 and look at different installation options, such as the classic and well-known *wipe-and-load* option, the frequently used *in-place upgrade* option, and the more modern *Windows Autopilot* option.

This chapter demonstrates the conditions under which an in-place upgrade is possible. It will explain the new Modern Lifecycle Policy of Windows 11 and what effect it has on the older Windows 10. We will provide decision support for choosing the right channel (annual or LTSC). Additionally, we will show which activation options are available in an enterprise environment. We will round this chapter off with tips and tricks for a smooth in-place upgrade.

In this chapter, we will cover the following topics:

- Differences between the annual channel and Long-Term Servicing Channel
- Risks and support lifecycles of the channels
- Hardware requirements for Windows 11
- Deployment methods: in-place upgrade, provisioning, and Autopilot
- Limitations and blockers of an in-place upgrade
- Problems with the traditional wipe-and-load method
- Activation options in an enterprise environment
- Tips and tricks for a smooth in-place upgrade from 8.1, 10, or 11 [21H2] to the latest 11 [22H2]
- Selecting the correct deployment tool

Selecting the edition and channel version

Windows 11 is available in different **stock-keeping units** (**SKUs**) (also known as editions) besides the **Home** edition, which doesn't play a role in the professional environment. Other available editions include **Pro**, **Pro for Workstation**, **Education**, **Enterprise**, and **Enterprise LTSC**. For business use, you should go with Enterprise or Pro/Pro for Workstation, depending on your licensing.

There are also other special editions, such as the **Team** edition, which is installed on the Surface Hub, the **Holographic** edition, which runs on Microsoft Hololens, and the **IoT Enterprise** edition, which is a variation of Enterprise LTSC in terms of licensing, but not in terms of bits and bytes.

In addition, there is a new **SE for Education** version, which is kind of "Windows 11 light." It has a reduced hardware floor for cost-effective devices in education, as well as SKUs such as **Multi-Session**, other special SKUs, and regional variants (N/KN/China); however, these are not within the scope of this book.

An explicit Windows 11 S edition is no longer offered. **S mode** is only available in the Home edition as an option. If you have Windows 10 installed, for example, if you're in Pro with S mode and want to do an in-place upgrade, you must first exit S mode and then see the *Limitations and blockers of an in-place upgrade* section of this chapter.

> **Important note for enterprise customers**
> There are no special licensing requirements for Windows 11 beyond those for Windows 10 devices.

Microsoft 365 licenses that include Windows 10 licenses allow you to run Windows 11 on supported devices. If you have a volume license, it covers Windows 11 and Windows 10 devices equally, before and after the upgrade.

Home users can currently upgrade from Windows 10 to Windows 11 for free.

General Availability Channel (GAC) and support timeline

Windows 10's new "Windows-as-a-service" concept is continued with Windows 11, and you can choose between two main flavors. All Home, Pro, Pro for Workstation, Enterprise, and Education SKUs are available in the regular updating channel, which is now called the **General Availability Channel** (**GAC**).

In addition, the **Long-Term Servicing Branch** (**LTSB**) was renamed the **Long-Term Servicing Channel** (**LTSC**) in 2018. LTSC is only available for the Enterprise SKU. More details on the LTSC are in the next section.

In 2015, the early days of Windows 10, the regular updating version was known as the **Current Branch** (**CB**) and in 2017 was renamed the **Semi-Annual Channel** (**SAC**). For a while, there was also the **Current Branch for Business** (**CBB**), which was supposed to symbolize the status of *Enterprise Ready* for the current Windows 10 edition. The name CBB was retired without replacement a long time ago.

Also, the term Semi-Annual Channel (Targeted), which was supposed to help with piloting, has disappeared in the meantime.

With the release of Windows 11, the release cadence was decreased from semi-annually to annually. Therefore, the Spring issue was canceled. For internal and logistical reasons, the decision was made to release in the second half of the year in the future (aka H2). Until further notice, a new version will be published annually, and this will take place in the second half of the year.

In addition to this change, the support periods have now been increased from 18 (Consumer edition and Spring edition for corporate customers) and 30 (Fall edition for corporate customers) months to 24 (consumer) and 36 (enterprise) months, respectively. In respect of these changes, the previously named Annual Channel is now known as the **General Availability Channel** (**GAC**), which is the channel used for software updates.

We are pleased that Windows 10 will begin with 21H2 and only be released in an annual cadence from now on. These decisions help enterprise customers with the transition to Windows 11. Future annual Windows 10 versions are also planned for the second half of the year; however, the support period for Windows 10 will not change.

The first Windows 11 version was technically Windows 11, version 21H2, and was released in October 2021, carrying the build number 22000. Build numbers are currently the only way to clearly differentiate Windows 10 and Windows 11 and distinguish the different yearly releases from one another. We'll explore this more in a later section.

To have less confusion between Windows 10, version 21H2 and Windows 11, version 21H2, it was decided to market the latter as simply *Windows 11* in the beginning. Microsoft then changed the name of the release back to Windows 11, version 21H2 with the release of Windows 11, version 22H2 to enable differentiation between the two versions.

Windows 11, version 22H2 is the second version of Windows 11 and was released in September 2022. The build number of the latest edition is 226xx. If you come across a build number of 25xxx or greater, these are the first Insider builds of the 2024 (or later version) development branch.

This annual channel is the recommended channel for most enterprise customers and standard Office PCs. Office 365 is fully supported on the annual channel.

In the annual channel model, the system is updated yearly. As soon as a new version is available, it will be rolled out to all Windows 11 installations, which will get their updates directly from **Windows Update** (**WU**) or **Windows Update for Business** (**WufB**) online. The rollout will be done in stacked waves.

If you want to postpone such a rollout, you need to defer feature updates, which is an option only available in Pro, Pro for Workstation, Enterprise, and Education. You can defer updates per **Group Policy Object** (**GPO**) (**Windows Components | Windows Update | Manage updates offered from Windows Update**) when using WU for up to 365 days.

Please refer to GPO's *Select the target Feature Update version* and *Select when Preview Builds and Feature Updates are received.* (See `https://packt.link/sY5tr` and `https://packt.link/G5EIm`.)

With **Microsoft Intune Modern Device Management** (**MDM**), you have more granular settings under **Devices | Windows | Feature updates for Windows 10 and later**. In addition to the available **ASAP** option, there is also the possibility to define a specific time for the global rollout as well as a gradual rollout with a start date, end date, and the number of days between these auto-created groups.

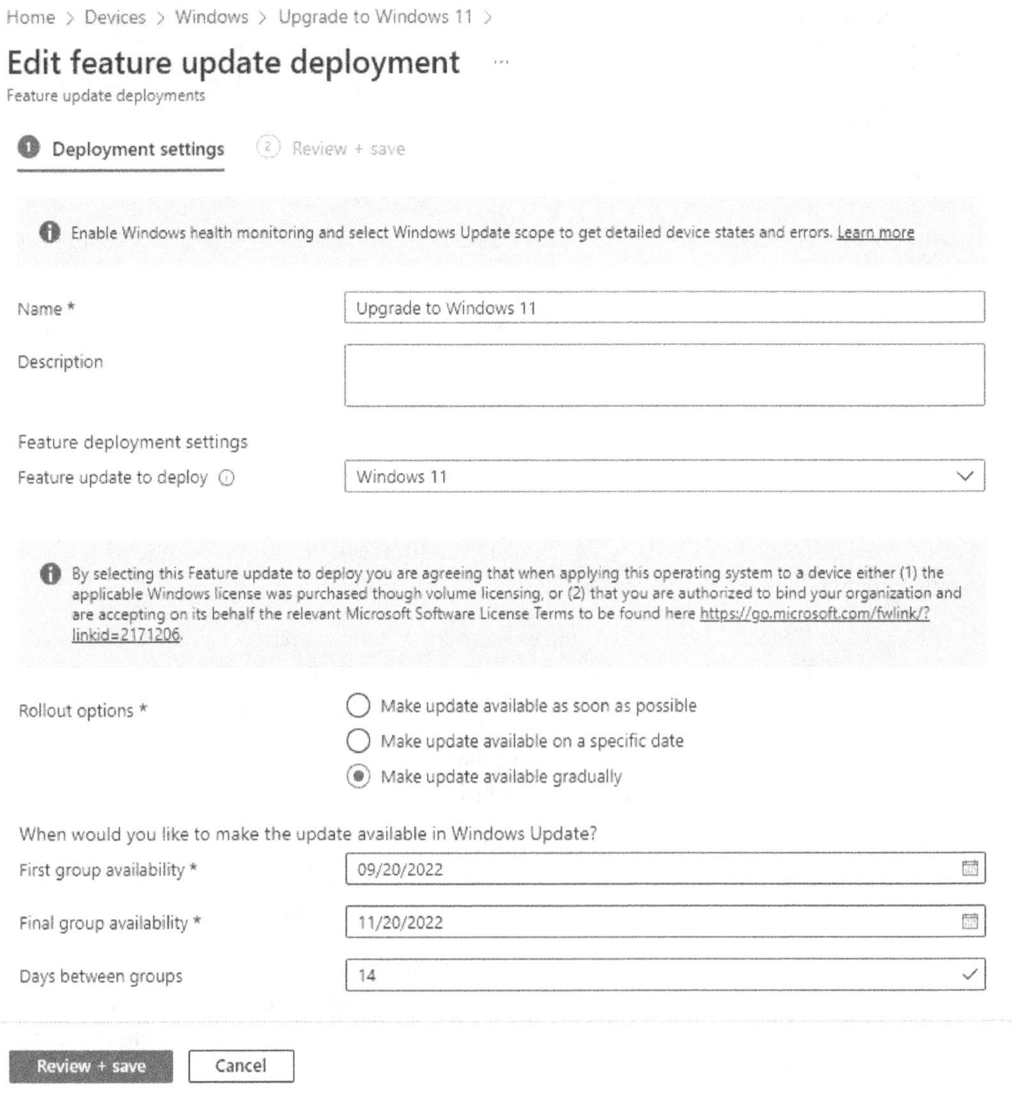

Home > Devices > Windows > Upgrade to Windows 11 >

Edit feature update deployment   ···

Feature update deployments

1 Deployment settings   ② Review + save

ⓘ Enable Windows health monitoring and select Windows Update scope to get detailed device states and errors. Learn more

Name *

> Upgrade to Windows 11

Description

Feature deployment settings

Feature update to deploy ⓘ

> Windows 11   ∨

ⓘ By selecting this Feature update to deploy you are agreeing that when applying this operating system to a device either (1) the applicable Windows license was purchased though volume licensing, or (2) that you are authorized to bind your organization and are accepting on its behalf the relevant Microsoft Software License Terms to be found here https://go.microsoft.com/fwlink/?linkid=2171206.

Rollout options *

◯ Make update available as soon as possible
◯ Make update available on a specific date
◉ Make update available gradually

When would you like to make the update available in Windows Update?

First group availability *

> 09/20/2022   📅

Final group availability *

> 11/20/2022   📅

Days between groups

> 14   ✓

Review + save   Cancel

Figure 1.1 – Intune feature update deployment settings

A new cloud service called Windows Autopatch is now generally available. It automates Windows, Microsoft 365 Apps for enterprise, Microsoft Edge, and Microsoft Teams updates, and can help with automatically patching the devices in waves driven by automatic issue detection for pausing and rollback. Learn more about this new option at `https://packt.link/SWVXs`.

In on-premises environments, you can directly defer feature updates inside your **Windows Server Update Service (WSUS)**, **Microsoft Endpoint Configuration Manager (MECM aka SCCM)**, or third-party deployment solution for an even longer time frame. For more information, see *Chapter 6, Device Management*.

Long-Term Servicing Channel (LTSC) and support timeline

Since 2021, the **Long-Term Servicing Channel (LTSC)** has only had a five-year support time frame, which is the same time frame as former Windows releases. During this five-year time frame, the LTSC will get security and quality updates, but no feature updates. Stability and not breaking anything are the critical focus points of updates during this time frame.

LTSC versions are only available as Windows 10/11 Enterprise LTSCs. So, if you do not have Windows 10 or 11 Enterprise, you won't qualify for LTSC. The version always contains a year in its name. LTSB/LTSC versions are referenced as Windows 10 Enterprise LTSB 2015, LTSB 2016, and so on. Since the change to LTSC, they are known as LTSC 2019, with the latest version being LTSC 2021. New LTSC releases are planned typically every two or three years.

To get new features, you will need to install a newer LTSC version. Microsoft never publishes feature updates through Windows Update on devices that run Windows 10/11 Enterprise LTSC.

No LTSC version for Windows 11 has been released yet. The last LTSC version was released in 2021 and is still a Windows 10 edition (technically corresponding to Windows 10 21H2). Upon release of the LTSC 2021 edition, the maximum support period was also shortened from 10 to 5 years, meaning that from now on, this and all future LTSC versions will only be supported for a maximum of five years. The former 2015, 2016, and 2019 releases will have 10 years of support.

According to an announcement via the Microsoft Tech Community, an LTSC version based on Windows 11 is not planned until 2024.

IT pros who are getting nervous when reading about the former two updates per year in the SAC now becoming once a year in the GAC may be tempted to select the LTSC as it looks like all the previous Windows versions' support strategies at first glance. However, there are several risks and limitations when choosing the LTSC.

The LTSC was designed for specialized systems such as controlling medical equipment, point-of-sale systems, and ATMs. These devices typically perform a single important task and don't need feature updates as frequently as other devices.

> **Microsoft's statement from the LTSC documentation**
>
> The LTSC is not intended for deployment on most or all of the devices in an organization; it should be used only for special-purpose devices. As a general guideline, a device with Microsoft Office installed is a general-purpose device, typically used by an information worker, and therefore it is better suited for the GAC (https://packt.link/WHc6o).

Maximum compatibility, reliability, and stability are the key focuses of the LTSC, which makes changes to the kernel and system less possible. Using MS Office and other products on your system that require changes to the system would block a patch. Therefore, you could end up in a situation where the only workaround would be waiting for the next (fixed) LTSC or changing to the GAC in the meantime.

In the LTSC version, some programs (modern apps) are removed or replaced by the older legacy Win32 apps (for example, calculator, notepad, etc. are replaced, and Edge and others are missing) compared to a SAC/GAC version. In the LTSC version, modern apps (appx) are not supported for sideloading (even if technically possible). Official note from Microsoft to this topic:

> **Microsoft's note on the support of APPX on the LTSC**
>
> The LTSC is available only in the Windows 10 Enterprise LTSC editions. This edition of Windows doesn't include a number of applications, such as Microsoft Edge (classic), Microsoft Store, Cortana (though limited search capabilities remain available), Microsoft Mail, Calendar, OneNote, Weather, News, Sports, Money, Photos, Camera, Music, and Clock. These apps are not supported in the Enterprise LTSC editions, even if you install using sideloading. (https://packt.link/jEDVo.)

In 2020, Microsoft also updated its Office 365 system requirements, and as of January 2020, Office 365 is no longer supported on any Windows 10/11 LTSC/LTSB release. If you plan to use an LTSC version, you will need to also use an LTSC version of Office (for example, Office 2021 LTSC). An explanation of this topic can be found at following Microsoft article: https://packt.link/PbAFR.

Silicon support policy and the LTSC problem – a potential risk with CPU availability and newer CPU generations

Besides the matter of app support, there are also important things to note on the hardware compatibility of future CPU generations in the LTSC.

> **Microsoft's note on LTSC CPU support**
>
> LTSC releases will support the currently released processors and chipsets at the time of the release of the LTSC. As future CPU generations are released, support will be created through future LTSC releases that customers can deploy for those systems. For more information, see *Supporting the latest processor and chipsets on Windows* in *Lifecycle support policy FAQ - Windows Products*: https://packt.link/pcVgo and https://packt.link/oWgse.

At the time of the LTSC 2021 release, the latest processor families in 2021 were Intel's Alder Lake (12th Gen) and AMD's Zen 3 platforms. Newly released processors such as AMD Zen 4 or Intel Raptor Lake (13th Gen) are not guaranteed to be supported on LTSC 2021 as they may need modifications to the kernel and the system, and this conflicts with the maximum reliability and compatibility goals. Each new CPU generation will be decided on a case-by-case basis depending on the changes needed to the system. The decision will be communicated as soon as the CPU generations are officially released.

So, even if AMD's Zen 4 and Intel's Raptor Lake get support in LTSC 2021, you can end up with the next CPU Generation Zen4+ or Meteor Lake (14th Gen) without support in your LTSC 2021 and may need to wait for LTSC 2024 or switch to the GAC version. This will significantly impact your five-year usability of the LTSC.

The supported CPU generations for each Windows version are documented at `https://packt.link/cmIWa`.

Further limitations of the LTSC

The LTSC has some more important limitations:

- **Limited in-place upgrade support**: Since 1607/LTSB 2016, an in-place upgrade from LTSB/LTSC to an equivalent or newer SAC/GAC is supported. However, an in-place upgrade from a previous OS (Win 7/8.1) to LTSC, or the change from SAC/GAC to LTSC is still not supported and is not planned for the future.

- **No suitable hardware support**: You may also find yourself in a situation where the CPU generations supported by the current LTSC version are no longer available as new devices, but the future LTSC version with the appropriate support for these CPU generations has not been released either, so you have an image but no suitable hardware.

Although it would be possible to circumvent these limitations by procuring sufficient supported hardware and stockpiling replacement devices, this would involve additional costs. Also, people might now say lightly that they don't need the in-place upgrade anyway. However, I have already seen several companies maneuver themselves into dead ends because they suddenly needed this feature and could only perform a time-consuming wipe-and-load. This leads us to our LTSC deployment recommendations.

Recommendations

With all the limitations and caveats of LTSC, it is best to stay with the GAC for most of your PCs. Use the LTSC only in situations where long-term maintenance is essential, such as in production lines, point-of-sale systems, and medical control systems. Most enterprise customers decide to roll out the GAC on their main general-purpose systems, and so should you.

Hardware requirements for Windows 11

In many ways, Windows 11 represents an innovation, coming six years after the release of Windows 10. Not only is it a major release for the first time since Windows 8.1, which causes support for several older CPU generations to expire, but it also represents a milestone on the client level with the end of 32-bit support. What we are already used to on the server side beginning with Server 2008 (Server 2008 was the first server OS offered as a 64-bit version only) will now also become standard for client operating systems. Windows 11 and all future versions will only be released as 64-bit (also on the ARM side). More details about the reasons for this CPU decision can be found in the *CPU limitations* section further on in this chapter.

Official (minimum) requirements

There are other important system requirements that can also be a stumbling block, such as **UEFI**, **TPM 2.0**, and so on. Therefore, let's take a closer look at the minimum requirements from https://packt.link/KJFRa.

Processor	1 gigahertz (GHz) or faster with two or more cores on *a compatible 64-bit processor* or **System on a Chip (SoC)**.
RAM	4 GB.
Storage	64 GB or larger storage device.
System firmware	*UEFI*, Secure Boot capable.
TPM	**Trusted Platform Module** (TPM) version 2.0 or Pluton Security Module.
Graphics card	Compatible with DirectX 12 or later with a WDDM 2.0 driver.
Display	High-definition (720p) display that is greater than 9" diagonally, 8 bits per color channel.
Internet connection	Windows 11 Home edition requires internet connectivity and a Microsoft account. For all Windows 11 editions, internet access is required to perform updates and to download and take advantage of some features. A Microsoft account is required for some features.
S mode support	S mode is supported only by the Windows 11 Home edition. If you are running another edition of Windows 10 in S mode, you must first switch out of S mode before upgrading to 11. Switching a device out of Windows 10 in S mode also requires an internet connection.

Table 1.1 – Minimum Windows 11 requirements

Let's walk through these values together and give complimentary notes on each.

We have dedicated a separate section called *CPU limitations* to the topic of Windows 11 processors.

Windows 11 requires at least 4 GB of memory, but this is the minimum required only for basic functions. Application programs can have higher requirements, but Windows 11 functions also have significantly higher requirements. For example, Application Guard for Edge requires at least 8 GB of RAM. If you run several application programs, you will very quickly fall below the minimum RAM requirement with only 8 GB RAM to start Application Guard for Edge. If you use Application Guard for Office, you should instead plan for 16 GB in order to be able to run this performantly, as a 600-800 MB virtualization container and the virtualized Office output are also executed on the host operating system besides Windows 11 and Office in this case.

The 64 GB storage represents a doubling of the previous Windows 10 specifications of 32 GB. The OS has not grown significantly, but we have noticed that with *only* 32 GB storage, problems with insufficient memory for the in-place upgrade occur disproportionately often. Even the specified 64 GB is still very ambitious, and you should not install too much extra on this OS partition or activate storage-heavy Windows 11 functions such as Application Guard Container, Windows Subsystem for Linux, Windows Subsystem for Android, Windows Sandbox, or similar. Therefore, 128 GB should be the minimum storage requirement, and we recommend 256 GB or 512 GB for Office PCs. More about this is coming up in the *Recommendations for a future-oriented hardware choice* section.

Windows 11 requires a UEFI firmware that (if still switchable between legacy mode and UEFI mode) runs in pure UEFI mode. Note that legacy mode is often referred to as BIOS mode or CSM mode). The UEFI mode must also support Secure Boot, and Secure Boot needs to be enabled. This disqualifies early UEFI implementations from the Windows 7 era, as UEFI 2.3.1 Standard or higher is required, which premiered with Windows 8. Because Windows 7 was not 100% compatible with UEFI **Graphics Output Mode (GOP)** (you could install Windows 7 on such UEFI 2.3.1 computers in UEFI mode, maybe you still got the startup sound but then had a black screen), and for a supported in-place upgrade, UEFI must be on, so an in-place upgrade from Windows 7 to 11 directly is out of the question. We will describe a possible workaround in the *In-place upgrades* section.

Windows 10 already required **Trusted Platform Module (TPM)** 1.2 as a minimum for many security functions, but TPM 2.0 was already required for some important functions to be able to activate these functions. However, these security features were not mandatory and **Original Equipment Manufacturers (OEMs)** still had the option to offer Windows 10-compatible logo devices without TPM. Since the cyber threat situation is getting worse, and Microsoft has moved the goals of Zero Trust and security from the chip to the cloud, it is only logical to activate some of these important security features by default for new installations and to require TPM 2.0 for them.

There is an attack vector where the communication between the CPU and the dedicated TPM chip can be eavesdropped on. In the past, TPM firmware updates were sometimes slow to be rolled out by the OEMs, or older devices were completely forgotten by them. Therefore, Microsoft has addressed these weak points with the optional Pluton security chip, which is integrated directly into the CPU and can receive firmware updates directly from Microsoft. Find out more about the Pluton chip in *Chapter 8*, *Windows 11 Security*, and at *Meet the Microsoft Pluton processor – The security chip designed for the future of Windows PCs* (`https://packt.link/jtBtx`)

The requirements for graphics cards and minimum display resolution should be fulfilled by current hardware. Certain Windows 11 functions, such as 3-column snap, require at least 1,920 pixels. There will be more about this in the *Hardware requirements for additional features* section.

The internet requirement and the necessary Microsoft account only apply to consumer editions. However, some Windows 11 functions may require internet access or have limited functionality without the internet.

There are several ways to remove the limitations of CPU and TPM compatibility. Besides various tinkering solutions that modify the setup files, intervene live in the setup process, or perform other unrecommended hacks, there is also the possibility to remove these limitations via a registry key. In the meantime, these registry keys have been officially documented by Microsoft. (`AllowUpgradesWithUnsupportedTPMOrCPU`). However, Microsoft points out clearly that with circumvention of the minimum requirements, you take on the risk of possible limited functionality, crashes, instabilities, and other problems. In addition, such systems do not receive any support from Microsoft, and it is not guaranteed that future Windows 11 updates can be installed.

Therefore, *it is strongly advised not to override these minimum system requirements on production systems*. This option is only for a test lab scenario at most. Computers that do not meet the minimum requirements will also be watermarked in the future. Computers that do not meet the minimum requirements of Windows 11 can continue to run Windows 10 until October 2025.

CPU limitations

It gets a bit more complicated with the CPU since there is a dedicated compatibility list as well as the basic requirements of a minimum of 1 GHz, 64-bit compatibility, and two cores.

Currently, this list roughly includes all Intel CPUs beginning with the 8000 series (Coffee Lake) and similar, as well as AMD CPUs beginning with the Ryzen 2000 series (ZEN 2) and similar. The list has been extended for the Intel 7000 generation. The X and W Xeons and 7820HQ variants are now also allowed, but only if *modern managed* drivers are used. That means only **Declarative Componentized Hardware (DCH)**-supported app drivers are used, for example, in the Surface Studio 2.

An extension of this list to older CPUs is currently not under discussion.

The most up-to-date lists for each client OS can be found under *Windows Processor Requirements*: Intel processors supported under Windows 11 | Microsoft Docs: `https://packt.link/P5loz`.

There are dedicated lists of supported Intel, AMD, and Qualcomm processors for Windows 11 [21H2] and Windows 11 22H2, respectively (as well as the older client OS and server OS). As soon as new versions of Windows 10 and 11 are released, this list will be updated.

Unfortunately, this limitation of the CPUs was communicated to the public very late. But how did this CPU selection come about?

Here are some of the reasons and decisions that led to this list. Among other things, the Microsoft team conducted evaluations around crashes and reliability. It was noticed that older systems had significantly higher crash rates than newer systems. Whether this was caused by outdated BIOS versions, the subsequent changes by Spectre and Meltdown mitigations, older/poorly maintained OEM drivers, or other reasons, was not specified.

In addition to ensuring system stability, system performance emerged as another crucial factor. Windows 11 mandates the default activation of several security features, notably **Virtualization-Based Security (VBS)** along with **Hypervisor-Protected Code Integrity (HVCI)**. This led to the collection of relevant telemetry data throughout the Insider Preview phase. It's noteworthy that HVCI, when executed without compatible hardware support, can result in a substantial slowdown of individual system calls, potentially ranging from 600% to 800%. The essential hardware compatibility required for optimal HVCI operation is provided by what's known as **Mode-Based Execution Control (MBEC)**. MBEC was initially introduced with Intel's seventh-generation Kaby Lake CPUs and AMD's Zen 2 CPUs. As a result, processors preceding this architecture, such as Intel's sixth-generation and AMD's Zen 1 generation, as well as earlier generations, lack the MBEC capability and were thus excluded from this feature. After careful examination of the accessible data, the decision was made to grant unrestricted support exclusively to Intel's eighth-generation processors. Additionally, support was extended to certain model series within the seventh generation, but only when utilizing DCH drivers.

As mentioned earlier, these specifications represent the bare minimum. To take full advantage of all the security features currently available in Windows 11, you'll need an Intel processor from the eleventh generation or newer, or an AMD processor from the Zen 3 generation or newer. Additionally, certain optional Windows 11 features might necessitate a more powerful CPU than what's specified as the minimum requirement.

Hardware requirements for additional features

The full list of additional features that we discuss here in some detail can be found at `https://packt.link/iz3yj`.

The list of additional features is very extensive, and many requirements are self-explanatory. You must have a Wi-Fi 6E-capable WLAN device installed to be able to use Wi-Fi 6E, or a 5G modem to be able to use the 5G functionality of Windows 11. It should also be obvious that you need to have an HDR-capable monitor to be able to use Auto HDR.

But some additional features do not have such clear hardware requirements. That's why we'll pick out a few of them that have requirements that are not immediately obvious:

- **Client Hyper-V** requires a processor with **second-level address translation (SLAT)** capabilities, activates the required features in UEFI firmware to use SLAT, and is only available in Windows Pro editions and above. (In some UEFI firmware, you need to activate VT-d/VT-x and sometimes also SR-IOV to have SLAT available. You can check SLAT availability via `coreinfo.exe -v` from Sysinternals Suite.)

- **DirectStorage** requires an NVMe SSD storage with a minimum of 1 TB and a minimum PCIe 3.0 connection, but a 4x PCIe 4.0 connection is recommended. Additionally, a DirectX 12 GPU with Shader Model 6.0 support is needed. To utilize the DirectStorage API, you need to use the Standard NVM Express Controller driver.

- **Snap** with three-column layouts requires a screen that is 1920 effective pixels or greater in width. If you are using more than 125% DPI, you will need an even larger resolution.

- **Windows Hello Face Recognition** requires a camera that supports near-infrared imaging.

- **Clear Voice** requires a studio microphone array installed by the OEM. The first device to support it is Microsoft Surface Laptop Studio together with Windows 11 22H2. Other manufacturers will follow soon. More details can be found in *Chapter 10, Windows 11 21H2 and 22H2 Changes*.

- **Windows Subsystem for Android™** currently requires 8 GB of RAM, an SSD or faster storage, and a supported processor (Intel® Core™ i3 8th Generation, AMD Ryzen™ 3000, Qualcomm® Snapdragon™ 8c, or above). Further updates about applicable system requirements will be communicated as the product is rolled out to select geographies.

This list is only a selection of additional features that require further hardware equipment, but I think you get a good impression of what we wanted to convey. More information about hardware requirements will be mentioned when we discuss additional features in *Chapter 10, Windows 11 21H2 and 22H2 Changes*, with the important changes to Windows 11.

Recommendations for a future-oriented hardware choice

I have already been asked by several customers to translate the partly cryptic hardware recommendations (which are often absolute minimum requirements) into realistic hardware recommendations for an Office PC that includes the activated security options of Windows 11.

This recommendation is based on the following Microsoft documentation as well as discussions with senior program managers, product managers, and architects:

- **Windows 11 requirements – What's new in Windows | Microsoft Docs**: https://packt.link/7oEtP

- **Windows Processor Requirements | Microsoft Docs**: https://packt.link/sHADs

- **Update on Windows 11 minimum system requirements | Windows Insider Blog**: https://packt.link/uGDSf

- For additional functions, **Windows 11 Specs and System Requirements | Microsoft**: https://packt.link/LhNYS

We've already recommended TPM 2.0 for Windows 10, but TPM 2.0 will now be required for Windows 11 (or alternatively the future Pluton security chip, since this is backward-compatible with TPM 2.0). We support both dedicated TPM 2.0 chips and **firmware TPMs** (**fTPM**) according to the 2.0 standard.

Windows 11 now requires 4 GB of RAM, but as described previously, this is only the absolute minimum. Since we actively use security techniques such as Virtual Based Security with optional Credential Guard (which runs another virtualized OS in the background), and others are recommended (for example, Application Guard), we would still recommend at least 16 GB of RAM or more.

We have discussed the minimum CPU requirements in detail in the corresponding section. From the security side, one should now consider having Intel's eleventh generation or AMD's ZEN 3 generation or newer for new purchases. From the performance side, usage has changed significantly, especially due to Covid and the new home office working style, with many video conferences. In addition, more and more tasks have changed over the years. Therefore, the advice to opt for processors with more cores (4 or more) and to avoid utilizing ultra-efficient Y CPU power classes still holds true for regular work devices when considering the minimum specifications for Windows 11 usage.

A new point from the past few months is that thermal management is becoming more and more important with newer processors, and passively cooled devices or devices with a small cooling solution quickly run into thermal throttling. Unfortunately, this is difficult to recognize on paper; identification is often only viable in the test lab. Therefore, new devices should go through a corresponding in-house test course, where it is explicitly ensured that the devices are also capable of video conferencing for a long time without throttling/jerking, or even switching off. On the other hand, the cooling solution should not involve too much obtrusive fan noise.

For storage, please go for SSD or NVMe. Specs require at least 64 GB as a minimum size, but since numerous programs are still installed, 256-512 GB is more realistic as a minimum. In order to benefit from the new, especially fast DirectStorage technology, NVMe with at least 1 TB is required.

Since password-free logins will become more and more important in the future, Windows Hello and the necessary biometric sensors should be considered. For comfort and security reasons, many customers choose Windows Hello Face Recognition, which requires a camera that supports near-infrared imaging (corresponding models are usually marked as Windows Hello cameras). You should also take a closer look at FIDO 2.0 devices.

The previously mentioned three-column snap functionality requires a screen resolution of at least 1,920 effective pixels wide.

The Secured Core PC concept, which started with Windows 10 and was created for increased security, is further supported and expanded.

Upgrading to Windows 11

With the introduction of Windows 10, there was a change to the installation mantra. Earlier, it was recommended you create a golden image and always perform a wipe-and-load sequence. Now, since Windows 10, it is recommended you perform an in-place upgrade, and this has not changed with Windows 11. The same techniques you are familiar with in Windows 10 can be used with Windows 11.

The various feature updates of Windows 10 have already used the mechanisms of the in-place update. Therefore, the transition from Windows 10 to Windows 11 feels more like another feature release and goes smoothly and without problems. Windows 11 has 99.7% application compatibility. Out of more than 1.2 million applications tested, just over 3,000 applications are not compatible as of Spring 2022. Therefore, an in-place upgrade is the way to go if you want to move from Windows 10 to Windows 11 quickly. Microsoft customers with an E3/E5 contract can get compatibility assistance when having compatibility issues with AppAssure. Learn more at `https://packt.link/GWaGP`.

There are some caveats when migrating from older systems and some limitations, but we will discuss them in the upcoming sections. We will also look at the different new possibilities introduced with Windows 11.

In-place upgrades

With the improvement of the Windows servicing stack, the possibilities of in-place upgrades became faster and more robust. In-place upgrades aren't the go-to solution but will do well in many scenarios. Performing an in-place upgrade will preserve all data, settings, apps, and drivers, so it will reduce the complexity of migration, the transfer of user profiles, and the (re-)installation of programs.

A big benefit of performing an in-place upgrade is 100% rollback in case of failure. With a classic wipe-and-load approach, if there is something wrong after installation, the user ends up with nothing. This puts a high time pressure on IT to solve the problem. Often, this pressure results in a fast workaround of reinstalling the client a second time and losing all data, settings, apps, and so on.

When something goes wrong during an in-place upgrade, the system will completely roll back to its original OS and the user will still be able to work with their client. This gives IT some time to review what went wrong and try the upgrade again later when they have a fix. Even after a successful in-place upgrade, IT can roll back to the old OS for 10 days if something is not working as expected.

The current in-place upgrade process is divided into four phases, with multiple reboots in between.

Technically, a completely new Windows is built, and afterward, the necessary drivers, then the apps, and finally, the selected settings are taken over from the old Windows. For several Windows 10 versions, as much as possible has been brought forward to the downlevel phase (in which you can theoretically continue to work) in order to keep the actual real downtime of phases 2-4 as short as possible and be back in a working state as quickly as possible.

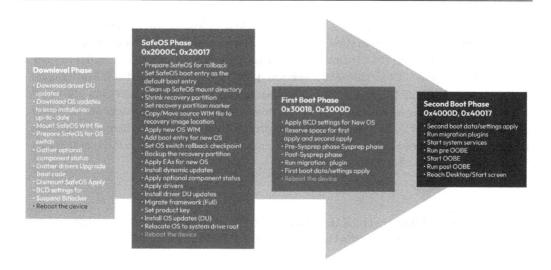

Figure 1.2 – Four phases of in-place upgrade

Let's look at the four phases in detail.

The downlevel phase

This phase is executed in the current OS. Depending on whether you are executing `setup.exe` or executing this phase by upgrading via Windows Update or WSUS, the GUI will be different or completely hidden. But technically, the following steps always need to be done:

1. Build a `$Windows.~BT` folder, analyze the system, and download required cumulative updates (if not restricted by setup flags).

2. Extract required drivers from the running system or (if not prohibited by setup flags) download drivers from Windows Update.

3. Prepare the system and the sources, place a SafeOS **Windows Preinstallation Environment (PE)** boot environment, upgrade the boot entry, and suspend BitLocker (if running).

Since Windows 10 20H1, you can specify whether you want to download the latest **Cumulative Update** (**CU**) for the target **operating system** (**OS**), any updated drivers available on WU, and/or any required updated **feature on demand** (**FOD**) and **language packages** (**LPs**). This makes it easier to work with multi-language and FOD scenarios in which you must always query the installed **Multilingual User Interface** (**MUI**)s and FODs in advance and then manually provide the necessary FODs and LPs in preparation for the installation. The system now recognizes which packages are necessary and fetches them from WU. This, however, requires the ability to connect to Microsoft endpoints to fetch the metadata/payload, something that many customers are preventing in their on-premises networks: `https://packt.link/euX7m`.

> Tip:
> With the release of Windows 11 22H1, the former divided ISO images for FODs and LPs are now combined into a single so-called **Languages and Optional Features** (**LOF**) ISO.

If the setup process encounters incompatibilities (so-called hard blocks as well as **SafeGuard holds**), it will terminate in this phase. In order to find out which app or which driver is responsible for the blockage or whether there is another reason for the blockage, you should take a closer look at the `CompatData XML` files in the `C:\$WINDOWS.~BT\Sources\Panther` directory. The evaluation of these logs can be simplified with the `SetupDiag.exe` tool. A hard block is permanent and must be solved to be removed.

On the other hand, there are so-called SafeGuard holds. These are temporary blockers that remain in place until a problem can be solved centrally. For example, a manufacturer could provide a bug-fixed driver via WU, an updated version of an app is available and is rolled out via app update/distributed by the manufacturer, and so on. As soon as the fix is provided by the manufacturer, SafeGuard holds are also removed after a certain safety time. All SafeGuard holds that occurred during the installation in the company can also be conveniently viewed in the Intune portal.

You will see this phase as **Windows Update preparing your system** on your screen, counting from 0% to 100%. The system will reboot after this phase. Setup result error codes (the second code after the `0xC19xxxxx` code) in this phase typically start with `0x100`.

The SafeOS phase

In this phase, a Windows **Preinstall Environment** (**PE**) instance is running, which is why it is called the SafeOS phase. The recovery partition will be prepared and updated, the old OS will move offline to `Windows.old`, and the new OS folder with the unpacked **WIM/ESD** (these are file extensions, which stand for **Windows Image file/Electronic Software Distribution**) that was built during the downlevel phase is moved to `C:\Windows`. Dynamic updates and OS updates will now be installed.

After that, the required drivers will be integrated so the system can boot from the new Windows version next time. You saw this phase in older Windows 10 releases as a black screen with a gray ring, like when doing a setup installation, and in releases since 1607 as a blue screen, like when installing Windows updates, with a message stating **Part 1 of 3** and counting from 0% to around 30%. This blue background is still the same in Windows 11. The system will reboot after this phase.

In the case of errors/rollback during this phase, take a closer look at the setupact.log file in the C:\$WINDOWS.~BT\Sources\Panther directory.

Setup result error codes (the second code after the 0xC19xxxxx code) in this phase typically start with 0x2000C or 0x20017.

The first boot phase

Now the new system will boot up for the first time and run through the sysprep phase. Device drivers are being prepared and the migration plugin is running to extract all remaining required data from the old OS. Already, the first boot data and settings have been applied. You will see this phase as **Part 2 of 3** and counting from 30% to 75%. The system will reboot after this phase.

If there is a crash/freeze/rollback in the range of 30% to 45%, this indicates problems with drivers. Sometimes setupact.log is not informative enough and you should additionally investigate setupapi.dev.log. It's especially important to check this if a section in this file aborts abruptly and is not written cleanly to the end (for example, there is an *installing driver* section but no comment on whether it was successfully completed). This may indicate that the last driver being migrated caused the setup to crash.

Setup result error codes (the second code after the 0xC19xxxxx code) in this phase typically start with 0x30018 or 0x3000D.

The second boot phase

In this last phase, the system runs the migration plugin one last time, applies the last migrated settings, starts the system services, and runs the **Out-of-Box Experience** (**OOBE**) phase. You will see this phase as **Part 3 of 3** and counting from 75% to 100%. The system will present you with the login screen after this phase.

Setup result error codes (the second code after the 0xC19xxxxx code) in this phase typically start with 0x4000D or 0x40017.

A very detailed TechNet article for information and guidance on various setup error codes and the possible locations of the log file during the different phases can be found at https://packt.link/D3Tuq. This link is valid for both Windows 10 and Windows 11 systems.

> **Which WIM/ESD files should I use for in-place upgrades?**
>
> The in-place upgrade process can only exchange your OS files. Therefore, it is only allowed to use a base ISO/WIM including only the operating system. The use of a sysprepped or so-called "golden system" image with additional apps included in the WIM is not supported. Microsoft could have designed a way around it to handle other Microsoft products such as Office, but this would end up in a two-class citizen system. So, the decision was made to only support the pure base OS upgrade.
>
> Only use an original WIM from `my.VisualStudio.com` or the **Volume Licensing Service Center** (**VLSC**). As a service, these ISOs on Visual Studio and VLSC are regularly updated monthly by Microsoft. They are usually available one week after Patch Tuesday also known as the 'B' release within Microsoft. It is allowed and recommended to maintain these WIMs offline with the latest cumulative update. To avoid unnecessary growth of the WIM file, start with the original WIM file each time.

With this move of Windows to `Windows.old`, the complete rebuild of the new OS, and only migrating the required settings, apps, and drivers, the whole process is now very robust and prevents too much waste from older OSs. Even so, Microsoft does not provide detailed numbers. We can only observe very low numbers of rollback (roughly below 2-3%) in huge national and international customers. And even if a rollback occurs, the system is still usable. Earlier Windows 10 versions had a slightly higher rollback of up to 5%, but it improved over time, and nowadays it is more likely below 2%.

This in-place upgrade process runs smoothly and quickly on modern hardware. We clocked approximately 15 minutes on a modern Intel i7 system with 512 GB NVMe and 60-70 apps installed. Even on a three- or four-year-old i5-class system with an SSD, times of approximately 25-30 minutes were possible. On very old hardware with low-RPM HDDs and a low amount of RAM, installation times can grow up to one and a half hours and more, but they would be on the lower end of the specs and should more likely be replaced than upgraded to Windows 11.

> **In-place upgrade as a repair option**
>
> An often unrecognized or underestimated option to repair unstable or broken systems is using an in-place upgrade to the same version. In other words, Windows 11 22H2 can be installed on a Windows 11 22H2 system again via an in-place upgrade. Since Windows is rebuilt and only settings/apps/documents are migrated, there is a good chance you will have a running and stable system again. It should also be much faster than a wipe-and-load and reinstall of all apps. The same applies of course to Windows 10 if you install, for example, Windows 10 21H2 on a Windows 10 21H2 system again.

Limitations and blockers of an in-place upgrade

With all the big benefits of an in-place upgrade, it still has some limitations and caveats. We will try to identify all blockers and limitations and formulate possible workarounds.

Changing from BIOS/legacy mode to UEFI mode

One problem is installations in BIOS or legacy mode. BIOS mode uses **Master Boot Record (MBR)** structure on your disk. Newer UEFI mode uses the modern **GUID Partition Table (GPT)** structure. You can perform an in-place upgrade in BIOS mode on Windows 10, but you will end up back in BIOS mode. Windows 11 now only supports UEFI, making a direct upgrade to Windows 11 impossible. Many enterprise customers wanted to take advantage of the new UEFI mode already under Windows 10, especially the new security features that require the secure boot feature, which in turn requires UEFI mode, so many systems should already be running in UEFI mode. Also, new devices in recent years were only allowed to have UEFI mode.

If you still have a Windows 8.1 or 10 system with legacy/BIOS mode that needs to be upgraded to Windows 11, you may want to do a classic wipe-and-load, considering all its problems such as losing settings, saving user data, and so on.

Since Windows 10 version 1703, there is another method supported by Microsoft to switch from BIOS mode to UEFI mode without reinstalling an operating system. The required tool, MBR2GPT.exe, is included in the operating system sources. MBR2GPT.exe must be run in offline mode, so run it from Windows PE or connect the hard disk as an additional disk to a running operating system and run it from there. This option is not available for LTSB 2015 and LTSB 2016. If you are already running a more recent Windows 10, you can convert MBR to GPT before upgrading to a new operating system. If you are working with a lower version of Windows 10, you need to upgrade to at least 1703 first. Of course, it makes sense to go to a version that is currently still supported, that is, 21H1 or higher.

The tool can only prepare the hard disk, convert the MBR to GPT, and update the boot entry. MBR2GPT. exe will not reconfigure the firmware. To change the firmware from BIOS to UEFI mode, you will have to use the manufacturer's tools yourself or do it manually. Since the tool interferes massively with the data structures of the hard disk, a previous backup is strongly recommended in order to avoid losing all data in the case of an error. However, if you make a backup, you can also re-partition the hard disk and restore the backup to the GPT partition.

Changing from Windows 32-bit/x86 to 64-bit/x64

A blocker that still exists and will persist in the future is trying to perform an in-place upgrade from an x86 to an x64 OS. In fact, you will be able to perform an in-place upgrade, and you will only be able to keep your documents, but not your apps and settings. There are too many changes with different paths, dual structures in the registry, and so on to keep this blocker going into future releases. Windows 10 was the last client OS to be released as 32-bit. With Windows 11, there is only 64-bit and a change is mandatory—you can keep your files or go with a fresh installation.

No direct in-place from Windows 7 to Windows 11

As mentioned previously, Windows 7 is not fully compatible with UEFI. In particular, **Graphics Output Protocol (GOP)** mode causes problems. Therefore, an existing Windows 7 UEFI installation is unlikely.

To get from Windows 7 to Windows 11 via an in-place upgrade, Windows 7 must already exist as a 64-bit installation. The hardware must also contain one of the few CPU configurations that are supported under both Windows 7 and Windows 11. If Windows 7 is available as a 32-bit installation, it is better to refrain from an in-place upgrade attempt, as there are usually too many problems with the changeover and a direct jump from 32-bit to 64-bit does not retain any apps or settings, so the added value of this type of in-place upgrade is very limited.

Since it is not possible to go directly from Windows 7 64-bit to Windows 11, the detour via an intermediate system must be the chosen method. If you switch from Windows 7 to Windows 8.1 first, the conversion from MBR to GPT must be done manually. If you switch from Windows 7 to Windows 10 right away, you can use the path described previously via MBR2GPT.

Changing the base OS language

You can't change the base OS language on the fly during an in-place upgrade. For example, if your current base OS is en-US with language packs installed for de-DE, you are not able to upgrade your system with a new base OS image with de-DE as the base language. If you try to do so, the in-place upgrade process will not successfully migrate your applications. There are some hacks on the internet that suggest installing all the language content of the new base language you want to change to, booting into Windows PE and setting the new international settings, and getting an OS that looks like it is specific to the changed language and will be accepted by the in-place update routine. But be warned: this is not officially supported (in terms of being tested by Microsoft), and so you could run into trouble and not be able to get official help.

Changing primary disk partitioning

Since Windows 10 1703, the partition type problem of MBR to GPT is solved, but all changes are still not possible out of the box. The included tools, such as `diskpart`, only support shrinking or expanding a partition, but not merging or moving a partition, or converting it from logical to extended, or vice versa. If you opted for a different disk layout involving these kinds of changes together with the new OS, you need to create the new structure before or after the upgrade with third-party tools, or go with the classic wipe-and-load option.

Using the Windows To Go or boot from VHD features

Since Windows 8.1, there have been no major improvements done to the **Windows To Go** feature; Windows To Go was deprecated in Windows 10. Since Windows 8.0, there have been problems performing an in-place upgrade, and these limitations still persist and will not be fixed. So, if you go with Windows 10 SAC in a Windows To Go scenario, you will end up doing a wipe-and-load once a year for your USB media. The only way to circumvent this problem is to use the LTSC version on Windows To Go media, but this comes with all the limitations and problems of LTSC. And as Windows To Go is deprecated, this is not an option when moving to Windows 11.

Boot from VHD, in terms of saving space, was replaced by a better feature called Compact OS. So, if you have only been doing boot from VHD for space-saving reasons, you should do a wipe-and-load and go with Compact OS in the future, which is fully in-place capable. If you've used boot from VHD for separating different (test) installations and want to use this feature in future versions, you need to do a wipe-and-load every time.

Image creation process (sysprep after upgrade not supported)

Before Windows 10, version 1607, you could not in-place upgrade your golden image and then do a sysprep. The process would detect such an in-place upgrade and present you with an error message: **Sysprep will not run on an upgraded OS**. This was changed with Windows 10, version 1607 and newer. Now this is a supported scenario: `https://packt.link/LPIZ3`.

The long version is still recommended, which is to always build your golden image from scratch with a base ISO of the new Windows version. As your running systems will be able to upgrade in-place, you will only need your golden image in the future for break fixes and installation of new hardware. In the case of new hardware, the process of reimaging will be replaced more and more by Autopilot.

Certain third-party disk encryption products

Even though there were various improvements in several Windows 10 releases, with `setup.exe` command lines to support more third-party encryption systems with the `/ReflectDrivers` parameter, you can still end up even with Windows 11 in a situation where an in-place upgrade is not possible or runs into a severe error due to your third-party disk encryption (driver). To prevent this problem, you need to upgrade to the newest version of your encryption product. If this still does not solve your problems during an in-place upgrade, you can only decrypt your drive (which is time-consuming) or use a classic wipe-and-load without retaining your apps, data, and settings.

Changing too many apps (bulk application swap)

Too much changing of applications is not a major blocker but could limit your installation times. As in-place upgrades only support the swapping of the base OS, all application changes need to be done before or after the upgrade inside task sequences. This could be very time-consuming, and when your top priority is deployment time, especially if there are no or low user data amounts and settings that need to be converted, the classic wipe-and-load could be a better option in terms of time taken.

Too many user profiles on the system

When there are too many user profiles and you install a feature upgrade, the `AppxMigration` plugin will possibly run into a timeout as it is a costly and a time-consuming process. We cannot give an exact hard limit at which it will fail, as it also depends on the amount of user data inside each profile. If an in-place upgrade fails due to timeout, try to remove unnecessary/no-longer-used user profiles.

> **Tip**
> Cleanup of unnecessary/no longer used user profiles before in-place migration will speed up the whole in-place upgrade process.

Changing the environment (change of domain)

Change of domain is not supported inside the in-place upgrade, as the corresponding sections inside your sysprep XML will be ignored. If you need to change domain, local admins, and so on, you need to do this before or after the upgrade. If it is not possible to run the new OS in the old environment or the old OS in the new environment, you could be limited to the classic wipe-and-load method.

Changing the environment (change from AD or AD/Hybrid to pure AAD)

It is not possible to upgrade from **Active Directory** (**AD**) or AD Hybrid to pure **Azure Active Directory** (**AAD**) as part of an in-place upgrade. Also, **User State Migration Tool** (**USMT**) does not support data migration between AD and AAD user accounts. By the way, it is also not possible to turn an AD Hybrid machine into a modern managed pure AAD machine by *just removing* AD, because as soon as you remove the AD config, critical configurations are also deleted and you cannot log in anymore. To get a pure AAD-managed modern client, you must either wipe-and-load or reset your existing Windows 10/11 and send it into the OOBE phase again to ensure a clean AAD join.

Traditional wipe-and-load

The well-known process of creating a golden image, sysprepping it, and deploying it to your clients (also known as wipe-and-load) will still be available with Windows 11 and still be supported in the near future. However, it is clear that more and more companies are moving to Autopilot and its associated benefits. So, you should also look into the Autopilot process, at least in the medium term.

To adapt your existing wipe-and-load process to the new Windows 11 OS, you just need to exchange your old **Windows Automated Installation Kit** (**WAIK**) or **Assessment and Deployment Kit** (**ADK**) with the newest ADK provided by the latest Windows 11 release. This implies that your deployment solution is capable of handling the newest ADK version. **Microsoft Endpoint Configuration Manager** (**MECM**) for example, requires MECM 2107 or later for Windows 11 21H2 and its ADK, and MECM 2207 or later for Windows 11 22H2 and its ADK. For other third-party deployment solutions, please consult their documentation or ask the vendor for their support requirements.

Please always use the appropriate ADK version for the OS, if possible. A newer ADK can be used with an older OS, but the reverse is not guaranteed and can cause problems. With Windows 11 22H2, the matching ADK version 22H2 has been released. For Windows 10, the latest ADK is version 2004, and (since Windows 10 only used enablement packages for newer feature releases from 2004 onwards) is still a valid alternative for Windows 10 2004 up to Windows 10 22H2. If a feature update is provided via an enablement package, the payload is already included in the monthly CUs and the enablement package just enables them. With Windows 11, versions 21H2 and 22H2 come as a full feature update with a swap of OS and so need a new ADK/SDK/DDK.

Carefully check every modification to the golden image to see whether it is still a supported setting or has been deprecated. Also check whether the feature you are (re-)enabling is perhaps now disabled by default (for example, SMB1 support). This could be a hint as to the expected support in future versions.

Do not modify binary registry keys, also known as binary blobs, and use only documented registry keys, because with every new version, undocumented registry keys can change behavior or vanish, even with cumulative updates. As they are not officially documented, there needs to be no official warning or announcement.

If you want to provide an experience such as in-place upgrades, you need more steps, more tools (for example, USMT), and possibly more external storage. In some rare cases, such as bulk app changes, you may need more time.

A down-level OS with lots of security mitigations, such as excessive use of `ICACLS.EXE` and other rights management tools to bend the security of the OS to comply with outdated or sub-optimally programmed applications, may also qualify for a wipe-and-load. If you use instead an in-place upgrade, you will possibly carry on these old changes and security risks which perhaps are no longer needed (or worst case now unsupported).

An alternative – provisioning package (PPKG)

Windows Configuration Designer (**WCD**) is part of the Windows 10/11 ADK and can also be found in the Microsoft App Store. It will be updated/enhanced with each release of the ADK. With the release of Windows 10 ADK 1703, it was renamed from **Windows Imaging and Configuration Designer** (**WICD**) to WCD. You should always use the newest WCD from the Store or the latest ADK.

WCD can not only create configuration packages, but it is also able to switch the SKU of your Windows 10/11 installation. You still cannot move to the LTSC via this mechanism, as this is a completely different build. At this stage of the process, you cannot downgrade. Currently, only an upgrade from Pro to Enterprise is possible (except for the Education SKU, which allows an upgrade from Home to Education).

Provisioning packages are very helpful when setting up kiosk mode. Also, you may find them useful when configuring the Microsoft HoloLens.

A good overview of all possibilities is provided at `https://packt.link/bd0Uc`.

WCD can be used to create packages that implement any MDM-based setting. Alternatively, you can run external scripts to set most MDM settings.

WCD has a wide range of functionality in addition to script support. All in all, it sounds like a mighty and powerful tool. However, it is currently not directly supported inside the Microsoft Deployment Toolkit GUI, MECM GUI, or Intune GUI. You need to wrap around a small PowerShell script and package it.

To force silent installation, you need to use the `Add-ProvisioningPackage` PowerShell cmdlet together with its `-ForceInstall` and `-QuietInstall` options and you should sign your package. You can also embed it in an image so that it gets installed during the OOBE process to get no prompts. However, this takes away a lot of flexibility.

> **Tip**
> If you hit the Windows key five times during OOBE, you can put in a provisioning package!

It has been our field experience that certain functions that WCD can perform or try will break the **Microsoft Deployment Toolkit (MDT)** and **Microsoft Endpoint Configuration Manager (MECM)** deployment process. Therefore, it should be tested first, and care should be taken when including this tool in your work.

WCD still lacks the ability to remove crapware and bloatware preinstalled onto vendor OS images. More information on WCD will be covered in the next chapter.

The modern way – Windows Autopilot

Windows Autopilot first appeared with Windows 10 1703 in 2017. Initially, Autopilot was not taken seriously nor seen as a real alternative due to the few configuration options and the requirement to use Intune/Azure AD. However, Windows Autopilot has grown with each feature release and now offers a wide range of options. It has also added good support for Hybrid Join and co-management for the transition period from OnPrem to Modern Managed AAD. In scenarios where many users share a computer, for example, call center systems, or computers that have to be set up as kiosk systems, Autopilot shows great strengths. But Autopilot also offers significant advantages when deploying outside the company network, which we now do more and more in the home office due to the Covid pandemic. Even post-Covid, we will certainly have to support the scenario to install a client for break-fix outside the company network more often. Therefore, if you have not already done so, you should really look into Autopilot.

Autopilot enables IT professionals to customize the OOBE for Windows 10/11 and enables end users to take a brand-new Windows 10/11 device and get a fully configured business device with just a few clicks. Users can walk through the self-service deployment of their new Windows 10/11 device without needing IT assistance.

IT will (optionally) pre-configure settings such as privacy settings, OEM registration, Cortana setup, OneDrive setup, choosing between personal or work devices, and preventing the account used to set up the device with local administrator permissions.

During Autopilot deployment, applications can be installed via MDM. Autopilot supports installation during the first login in the user-driven scenario, as well as automatic provisioning in the device-driven scenario. Additionally, the user-driven mode has the option to support the pre-installation of the apps by the service provider or the company's own IT before delivery to the user, through Autopilot pre-provisioning (formerly called Autopilot Whiteglove).

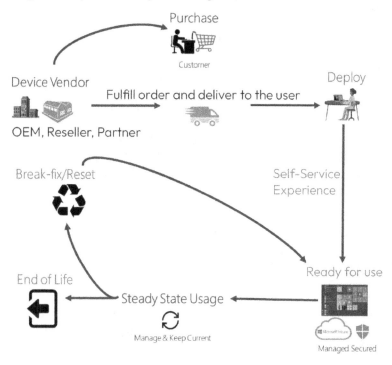

Figure 1.3 – Schematic flow of Autopilot

The device needs to be registered with your organization. IT will need to acquire the device hardware ID and register it. Microsoft is actively working with various hardware vendors and service providers to enable them to provide the required information directly to your organization via file or upload it to your tenant on your behalf.

When ordering new hardware in the future, you should make sure to get the necessary Autopilot information directly from the manufacturer, or alternatively have it uploaded to the tenant.

Some manufacturers also offer (for example, Microsoft with their Surface devices) to register the replacement device directly in the tenant again so that IT does not have to import this information again. This can be done if the device is already registered in Autopilot and a hardware replacement is necessary.

Autopilot device registration can be used together with Microsoft Store for Business or with Intune/third-party MDMs. We strongly recommend going with the Intune/MDM variant as the pure Store variant has too many limitations and will be retired in 2023.

For existing devices, the necessary information can be created with the help of a WindowsAutopilotInfo.ps1 PowerShell script. You can find this script at https://packt. link/DkYQz. Please use version 3.5 or newer.

The script can also be installed directly with the following PowerShell commands (needs admin rights):

```
Install-Script -Name Get-WindowsAutoPilotInfo -Force
Get-WindowsAutoPilotInfo -OutputFile AutoPilotInfo.csv
```

After that, you need to upload the CSV file obtained in this way in your Intune console.

If you have the appropriate Intune rights and want to upload this information directly during the first boot of the device, you can alternatively use the following commands during the OOBE phase (*Shift + F10* on the sign-in screen and after successful registration of the device, trigger a reboot to start the OOBE process again):

```
PowerShell.exe -ExecutionPolicy Bypass
Install-Script -name Get-WindowsAutopilotInfo -Force
Set-ExecutionPolicy -Scope Process -ExecutionPolicy RemoteSigned
Get-WindowsAutoPilotInfo -Online
```

If you use a current version of MECM or Intune, you can have this information automatically generated and uploaded from already managed systems. This can be used in Hybrid Join scenarios to register all devices in Autopilot. In order to switch from AD Hybrid Join to AAD only, the devices can then be reset and quickly brought back online via Autopilot OOBE.

The end user will unbox and turn on their new device. They just need to configure a few simple steps in the user-driven Autopilot scenario:

1. Select a language and keyboard layout.
2. Connect to the network.
3. Provide Azure AD email address and password.

All settings configured by IT will be skipped. Following this process, the device will be joined to Azure AD and enrolled into Microsoft Intune or another configured third-party MDM service. All assigned apps will be installed. Within a short time, the user has a working device with up-to-date drivers and Windows.

With the Windows Autopilot Reset capability, organizations can easily reset their configured devices while still maintaining MDM enrollment and the Azure AD join state to get the device back into a business-ready state very quickly.

> **Important note**
>
> An always up-to-date documentation page for Windows Autopilot, including the new features as soon as they are available, can be found at `https://packt.link/YVIYI`.

More details about Autopilot can be found in *Chapter 3*, *Configuration and Customization*, and *Chapter 9*, *Advanced Configuration*.

Activation of Windows 11

For corporate customers, there are three easy ways to activate Windows 11. The good news is that the same mechanisms can be used with Windows 10. Which of these options is the most suitable depends on whether you are running on-premises, hybrid joined, or AAD only. Let's take a closer look at the three options.

Classic activation by Multiple Activation Key (MAK) or Key Management Services (KMS)

In the on-premises and hybrid joined scenarios, you can still use the classic **Key Management Services** (**KMS**). If you use AAD only, only the **Multiple Activation Key** (**MAK**) can be used easily for the classic activation services.

KMS servers have the disadvantage of not offering any extra authentication and are therefore in need of special protection. They originate from an era of corporate LANs. Anyone who knows the IP address and listening port of the KMS server by reading the values of an active client can then use this information to activate any client on KMS. Or, even more simply, if KMS is registered in DNS, KMS can access the DNS server and query the information there without accessing an active client. KMS requires an extra port connection (TCP 1688). KMS is still supported and requires at least Server 2016 as the base operating system. But in an AAD-only scenario, you probably won't have a corporate LAN. How do you achieve KMS in an AAD-only scenario? Outsourcing KMS to the cloud requires careful planning to secure access to the KMS for the reasons we have just explored.

Alternatively, you can use a **Multiple Activation Key** (**MAK**). But a MAK is dangerous if it falls into the wrong hands. Each MAK also burns a license with every reinstallation.

More modern via Active Directory-based activation

Since Windows 8/Server 2012, there is an **Active Directory-Based Activation** (**ADBA**) service. This service offers more security advantages compared to the KMS standalone service. There is no need for an extra KMS TCP port because only the typical AD communication ports are used and there is also no need to run an extra server. However, authentication is required because only domain-joined computers are activated.

Even so, this last-mentioned advantage of authentication is also possibly a disadvantage, because workgroup computers cannot be activated. A former disadvantage of ADBA (that Windows 7 is not supported by this type of activation) should not play a role nowadays, because Windows 7 is only in **Extended Security Updates** (**ESU**) support until January 10, 2023 (with specially required MAK keys) and is also barely in use. This type of activation is only usable in on-premises and hybrid joined scenarios.

More information about this ADBA can be found at `https://packt.link/jkinA`.

Future-oriented via Windows 10/11 Subscription Activation

Since Windows 10 has the new option of **Subscription Activation**, the Enterprise license is directly assigned to an AAD user account. As soon as this AAD user logs on to a Windows 10 or Windows 11 system, it is immediately upgraded to Enterprise. Devices with a current Windows 10 Pro or Windows 11 Pro license will be seamlessly licensed to Windows 10 Enterprise or Windows 11 Enterprise accordingly. No upgrade from 10 to 11 will be performed.

If the device is already activated with an Enterprise key or KMS service, it can be easily switched to Subscription Activation. The big advantage of this activation type is that Pro images can be automatically *converted* to Enterprise images. So, it is no longer necessary to deploy an extra enterprise image on the device. Subscription Activation can be used in hybrid joined and AAD-only scenarios. This type of activation is not suitable for on-premises scenarios.

For the licensing to persist, the user must log in regularly every 30-90 days. If the user does not log in on this device for a longer period of time, it will fall back to the Pro license.

> **Important note**
> There are different licenses for human-operated systems and robotic systems.

There is one more thing to note when using **two-factor authentication** (**2FA**) and/or **Conditional access** (**CA**). In this case, please exclude the Universal Store Service APIs and Web Application, AppID `45a330b1-b1ec-4cc1-9161-9f03992aa49f` from all users all cloud apps MFA policy; otherwise, no automatic renewal of the license can take place.

Especially in AAD-only scenarios, the Subscription Activation method is absolutely recommended. More information about this new service is available at `https://packt.link/dMm3h`.

Tips and tricks for a smooth in-place upgrade from Windows 8.1 or 10 to Windows 11

The in-place upgrade is already very stable and robust, but with some tips, you can improve the robustness even more.

Looking up SetupDiag data in case of error/rollback

It is important to pay attention to the error codes and rollbacks and to look at the log files in the case of an error. This is the only way to detect a general problem at an early stage and provide a central solution quickly. `SetupDiag.exe`, which has been integrated directly into the sources since Windows 10 2004, is a very helpful tool. It automatically detects more than 40 possible causes of problems and provides an evaluation that is easy to read for humans. By regularly evaluating this data, scripted solutions can be developed quickly. The need for manual intervention therefore decreases significantly. If Microsoft Endpoint Manager Admin Center is in use, this data can also be viewed directly there.

Integrating cumulative updates into install sources

During the Insider Preview phase, several tens of thousands of different configurations will be tested, but some possible minor hiccups can remain in the very first ISO/WIM released directly at GA. If the automatic update function of `setup.exe` is not used, the image should be updated regularly to get all the improvements and fixes that need to be applied during an install/upgrade. Also, the list of hard blocks is updated with these integrated CUs.

The integration of CUs can be done by using the monthly updated versions from Visual Studio and VLSC, or the images can be updated manually. Upgrading your `install.wim` is very easy. Follow these steps:

1. Download the latest cumulative update from Windows Update Catalog.

2. Unpack the ISO and mount the included install.wim to a temporary folder.

3. Add the `.msu` file with `DISM.exe`

4. Commit the changes

5. Unmount the WIM file.

To reduce unnecessary growth of the WIM file, start over each time with the original WIM.

Updating drivers

In the past, graphics drivers were often problematic, especially if they were written for an older OS, but the situation has changed somewhat over the last two years. Due to increasingly strict security requirements (KMCI, NX+DEP, and so on) for drivers, crashes caused by sound drivers, network card drivers, or storage drivers are now also visible more often. Therefore, the recommendation is to keep all drivers as up to date as possible before upgrading.

Update the installed drivers of your down-level OS before attempting an in-place upgrade, especially if your driver is from before July 2015. Also update your SD card driver, as we've faced installations freezing several times during the first boot phase when initializing the SD card device. If there is still a problem in the 30%-60% first boot phase, try to detach unnecessary hardware during the upgrade.

Looking at Setupact.log and Setupapi.dev.log

`Setupact.log` and `Setupapi.dev.log` are perhaps the two most important log files that are used during update/setup failure troubleshooting. Here are the locations these files will be typically located, depending on the deployment phase:

- **Down-level** (`Setupact.log`):
  - **Usage**: Used for troubleshooting rollbacks and down-level failures
  - **Location**: `\$Windows.~BT\Sources\Panther`

- **Rollback** (`Setupact.log`):
  - **Usage**: Used to troubleshoot rollback and uninstall failures
  - **Location**: `\$Windows.~BT\Sources\Rollback`

- **Windeploy and OOBE** (`Setupact.log`):
  - **Usage**: Used to troubleshoot failures during OOBE
  - **Location**: `\$Windows.~BT\Sources\Panther\UnattendGC`

- **Pre-initialization** (`Setupact.log`):
  - **Usage**: Used to troubleshoot pre-launch failures
  - **Location**: `\Windows`

- **Upgrade Complete** (`Setupact.log`):
  - **Usage**: Used for post-upgrade investigations
  - **Location**: `\Windows\Panther`

This information should allow you to find and analyze the correct log file in all installation steps. Please always look at the log files for failed in-place upgrades to identify general problems (incompatible driver, blocking app, misconfiguration, and so on) as early as possible and fix them as soon as possible in order to have as smooth an upgrade process as possible.

Desktop Analytics and Microsoft Endpoint Manager admin center

The former service **Windows Upgrade Analytics** also known as **Windows Upgrade Readiness** was replaced in January 2020 by the new **Desktop Analytics** service. This service is available to Enterprise environments in conjunction with MECM, which makes use of the telemetry feature of Windows. While some view telemetry as a spying or data collection problem, Microsoft shows that they are using the data to improve Windows, while at the same time helping organizations upgrade to Windows 10/11.

The analytics features will work on Windows 8.1/10 and preceding hosts and allow the enterprise to gauge what hardware needs to be replaced before making the move to Windows 11. A detailed write-up of the service offered can be found in this article: `https://packt.link/9jU6J`.

The Desktop Analytics Service was retired on 30 November 2022. The information and insights currently found in Desktop Analytics are now directly incorporated into the **Microsoft Endpoint Manager Admin Center**. More information can be found in the *A data-driven approach to managing devices in your organization* blog at `https://packt.link/IkQ3C` and the *Windows 11 hardware readiness* section at `https://packt.link/i0upV`.

Selecting the deployment tools

It is not easy to select deployment tools. Different people will have different preferences and therefore favor different deployment tools. But perhaps we can roll up to the question from a different angle, as it is just the same if you use MDT, MECM, or a third-party deployment. You also need to decide whether you want to continue installing on-premises, or whether you are on the way to modern management with MDM and AAD (or have already arrived there). Accordingly, the deployment tool should be able to handle this option. MECM is well positioned for a transition with its co-management option.

It is important to use the latest ADK delivered with the Windows 11 release you are deploying. Your ADK should be one release older at the most, so have a look at the known issues page of the ADK before picking it. From this important requirement and the release cadence of one Windows 11 release per year, we come to the next prerequisite.

Your chosen deployment tool should get at least one to two updates per year to support the newest features and newest ADK. As more and more configuration is done not only by GPO, but also by MDM and the WMI bridge, it becomes more essential that your deployment and configuration environment keeps up with the pace.

Finally, the ability to script pre- and post-upgrade task sequences is important. You will most likely run into situations where you need to configure, update, or remove something before performing the upgrade, and the same applies to the phase after a successful upgrade. As some configurations can only be done from PowerShell, you should select a deployment tool with (direct) PowerShell support.

If you use the direct upgrade from MECM **Software Update Point** (**SUP**) for the feature upgrade, you should consider the possibility of using a task sequence for it. This possibility was recently integrated into MECM. The advantage of a task sequence used in connection with SUP is that you can do important configuration steps, clean-up work, and upgrades of existing applications promptly and before the actual upgrade, and can use a post-run/clean-up scripts after the upgrade. Unfortunately, there is currently no comparable task sequence option when using Intune and Windows Update for Business or **Windows Server Update Services** (**WSUS**) in standalone mode. Also, chaining between Intune packages and **Windows Update for Business** (**WUfB**) Feature Update as a replacement for a task sequence is currently not possible under Intune.

Summary

In this chapter, you learned the concepts and best practices of the available deployment options with Windows 11. We looked into the traditional wipe-and-load method and the complementary newer options of in-place upgrades, provisioning, and Autopilot, and provided some context to the difference these deployment options can make.

The next chapter will walk you through enterprise deployment and in-place upgrade techniques. Deployment tools will be covered, as well as tips and tricks to smooth in-place upgrades from Windows 8.1 or 10 up to Windows 11 and to migrate user state information and settings.

2
Introduction to PowerShell

PowerShell is a cross-platform (Windows, Linux, and macOS) task automation framework for IT professionals. PowerShell contains an interactive shell and a scripting language. The interactive shell is on par with the best Linux shells, including Bash. The scripting language is based on .NET and has the power of Perl, Ruby, and many other popular languages. PowerShell allows you to manage all aspects of Windows 11.

In this chapter, we cover the following:

- What is PowerShell?
- Installing Windows PowerShell and PowerShell 7 in Windows 11
- What are cmdlets, objects, and the pipeline?
- PowerShell's scripting language
- Modules and commands
- Hardening PowerShell
- Configuring PowerShell

It is not possible to describe PowerShell fully in a single chapter. This chapter aims to introduce the key concepts and terms, show examples, and give you resources to discover more about specific features.

What is PowerShell?

As mentioned, PowerShell is a task automation platform for IT professionals and comes to you in two forms: Windows PowerShell and open source PowerShell (aka PowerShell 7).

In this chapter, any material specific to either version of PowerShell is noted as Windows PowerShell or PowerShell 7. But for the most part, almost everything you know about Windows PowerShell is useful in PowerShell 7. However, there are some differences and improvements.

Windows PowerShell versus PowerShell 7

PowerShell comes to you in two broad forms: Windows PowerShell and open source PowerShell (aka PowerShell 7). Microsoft first shipped Windows PowerShell as an add-on to Windows XP and developed the product significantly in later releases. The concept of Windows PowerShell began with the Monad Manifesto (`https://packt.link/ZZHnW`). This document provides a fascinating insight into the development of Windows PowerShell.

After the release of Windows 8.1 and Windows PowerShell 5.1, Microsoft transitioned the product into the open source PowerShell 7. Microsoft also transitioned .NET Framework to open source (aka .NET), which underpins PowerShell 7. Windows PowerShell is feature-complete, and it is highly unlikely that new features will be developed. PowerShell 7, on the other hand, is under active development.

Windows 11 comes with Windows PowerShell 5.1 fully installed. You have to install PowerShell 7 if you wish to use it.

Microsoft developed Windows PowerShell on top of the Microsoft .NET Framework. In many cases, Windows PowerShell is just a thin wrapper around .NET. PowerShell cmdlets leverage .NET.

Microsoft moved .NET Framework and Windows PowerShell into open source (with Microsoft funding while running the development teams). .NET Framework became simply .NET and Windows PowerShell became PowerShell 7. If you read old documentation, you may find that the first versions of open source PowerShell were known as PowerShell Core and version 6.x. For the most part, these were proof-of-concept implementations and are no longer supported.

Why does this matter, you may ask? There are a few reasons. First, by moving to .NET, the development team did not port every component of .NET Framework. If you manage your system today using Windows PowerShell, some of your scripts may not work. One example is the WSUS module, which you would normally use to manage Windows system update services. The cmdlets in this module use some APIs that Microsoft did not make available in .NET, so the cmdlets do not work and there is no direct workaround. The WSUS product team need to re-engineer their module to make the cmdlets function with PowerShell 7. Fortunately, the number of modules that do not work is very low. The Windows compatibility feature, described in the next section, resolves most of these compatibility issues.

If you are familiar with Windows PowerShell, you should be able to use it immediately. *All* the language constructs in Windows PowerShell carry over into PowerShell 7. All the core PowerShell cmdlets (that is, those that come with Windows PowerShell) are available in PowerShell 7. This means that almost all your scripts should work fine. From the console, all the commands you use in Windows PowerShell work fine too. In most cases, if you can use Windows PowerShell, you know how to use PowerShell 7.

There are also some key differences between Windows PowerShell and PowerShell 7, most of which are improvements.

The main differences lie in the new and updated features contained in PowerShell. Each version of PowerShell 7 has incorporated large numbers of changes and improvements. To find the specific changes that each new version incorporates, see `https://packt.link/yPEPK`.

Windows PowerShell compatibility solution

A challenge for PowerShell 7 is that most of the Windows PowerShell modules, which Microsoft has published, do not work natively In PowerShell 7. To enable you to use these commands within PowerShell 7, the PowerShell team has developed the WindowsPowerShell compatibility solution.

This solution involves creating a PowerShell remoting session for the local machine (using a Windows PowerShell 5.1 endpoint), loading the module into the remoting session, and then using implicit remoting to create local functions that call the remote cmdlets.

This solution works very well – all but three Microsoft Windows Server modules work fine using the compatibility solution. One small issue is that the display XML that Windows PowerShell uses to format the cmdlet output is not imported with this solution by default. You can, however, manually load this XML.

Although the compatibility solution does not provide 100% fidelity, it gets close. And even for those modules that you cannot use directly in PowerShell 7, you can use PowerShell remoting as a workaround. The scripts at `https://packt.link/6w3FC` show how you can install, configure, and manage WSUS using PowerShell 7.

Installing PowerShell 7

Windows PowerShell comes built into every edition of Windows 11 (and the related versions of Windows Server). Microsoft also uses Microsoft Update to keep Windows PowerShell fully up to date.

Microsoft does not currently ship PowerShell 7 inside any version of Windows, although it is easy for you to install the product.

There are several ways you can install PowerShell 7 in Windows 11.

The first method is via the Microsoft Store. From Windows 11, you can use **Microsoft Store** and search for PowerShell 7 like this:

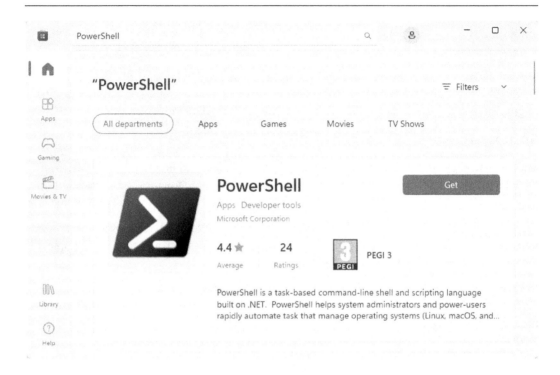

Figure 2.1 – Installing PowerShell 7 via the Microsoft Store

You click the **Get** button and follow the prompts to install PowerShell 7 via the Store.

The second method, and possibly the more IT-pro-friendly method, involves using the `Install-PowerShell.ps1` script, which you download directly from the PowerShell team's GitHub repository at `https://packt.link/C54Up`. This PowerShell script, created and maintained by the PowerShell team at Microsoft, allows you to install the latest released version using an MSI installer package that was created by the PowerShell team. You run this script, and it downloads and runs the MSI package (which also updates the Windows Path variable to enable Windows to find PowerShell post-installation).

Here is a simple script that downloads the latest released version of PowerShell 7 and installs it using an MSI file:

```
# 1. Download PowerShell 7 installation script
Set-Location C:\Foo
$URI = 'https://aka.ms/install-powershell.ps1'
Invoke-RestMethod -Uri $URI |
```

```
  Out-File -FilePath C:\Foo\Install-PowerShell.ps1
# 2. Install PowerShell 7
$EXTHT = @{
  UseMSI                   = $true
  Quiet                    = $true
  AddExplorerContextMenu = $true
  EnablePSRemoting         = $true
}
C:\Foo\Install-PowerShell.ps1 @EXTHT | Out-Null
# 3. For the Adventurous - install the preview and daily builds as
well
C:\Foo\Install-PowerShell.ps1 -Preview -Destination C:\PSPreview |
   Out-Null
C:\Foo\Install-PowerShell.ps1 -Daily    -Destination C:\PSDailyBuild |
   Out-Null
```

The output from running this snippet is as follows:

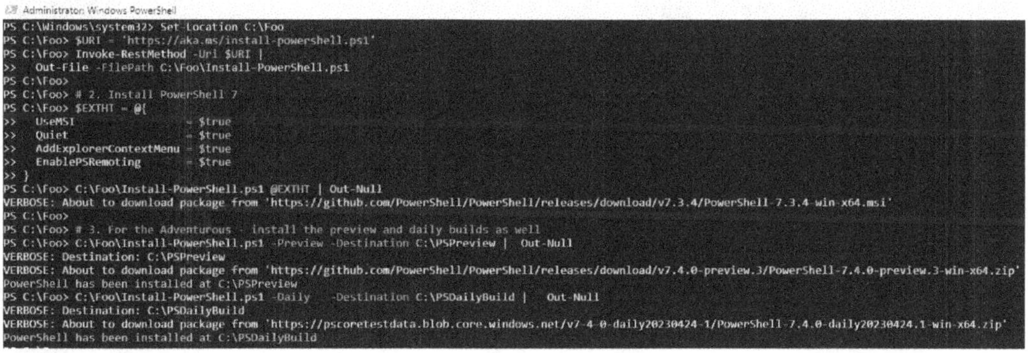

Figure 2.2 – Installing PowerShell

As seen in *Figure 2.2*, you can also use the `Install-PowerShell.ps1` script to download different versions of PowerShell to specific folders, enabling you to have multiple versions of PowerShell 7 installed, possibly for testing. For the brave, this script also allows you to install a preview build for the next version of PowerShell. For the ultra-brave, you can also use the latest build of PowerShell 7 (the build of the day), enabling you to evaluate some of the new features coming in the next release of PowerShell or added to the latest build.

Since PowerShell 7 is an open source project with a GitHub repository, you can download versions of PowerShell 7 directly from GitHub:

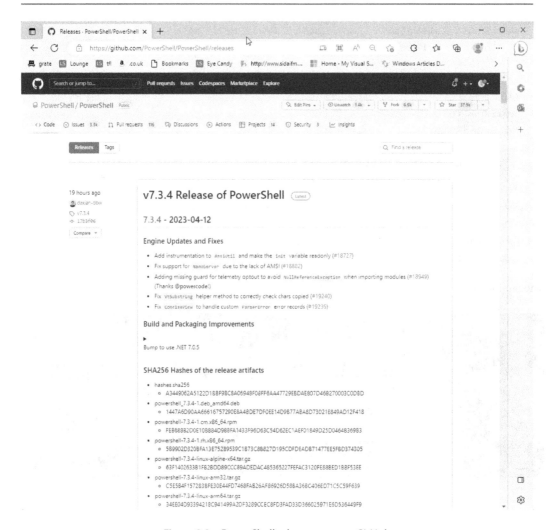

Figure 2.3 – PowerShell release page on GitHub

You can also use third-party package management tools, such as Chocolatey.

Keeping PowerShell up to date

Like almost every application, updates to PowerShell are a fact of life. With Windows PowerShell, Microsoft delivers updates via Microsoft Update.

Updating PowerShell 7 is a bit more complex, owing to the product's nature and the different installation methods available to you. Like so many things with PowerShell, you have options.

Whenever the PowerShell team releases a new version of PowerShell 7, they update metadata held on GitHub to indicate the latest version(s) available. You can see the metadata using the following snippet:

```
Function Get-PWSH7ReleaseInformation {
# Get details of overall PowerShell 7 information
  $FR = 'https://raw.githubusercontent.com/' +
  'PowerShell/PowerShell/master/tools/metadata.json'
  $MetaFullRelease = Invoke-RestMethod $FR
# Get Details of latest preview
  $MetaPreview = Invoke-RestMethod 'https://aka.ms/pwsh-buildinfo-
Preview'
# Get Details of the latest daily build
  $MetadataDaily = Invoke-RestMethod 'https://aka.ms/pwsh-buildinfo-
daily'
# Display this information
  'PowerShell 7 Status:'
  $MetaFullRelease
  'Preview information:'
  $MetaPreview
  'Daily Build information'
  $MetadataDaily
}
Get-PWSH7ReleaseInformation
```

When you run this code, you see something like this:

```
PS C:\Foo> Get-PWSH7ReleaseInformation
PowerShell 7 Status:

StableReleaseTag     : v7.3.4
PreviewReleaseTag    : v7.4.0-preview.3
ServicingReleaseTag  : v7.0.13
ReleaseTag           : v7.3.4
LTSReleaseTag        : {v7.2.11}
NextReleaseTag       : v7.4.0-preview.4
LTSRelease           : @{Latest=False; Package=False}
StableRelease        : @{Latest=False; Package=False}

Preview information:
ReleaseDate : 20/04/2023 18:04:01
BlobName    : v7-4-0-preview-3
ReleaseTag  : v7.4.0-preview.3

Daily Build information
ReleaseDate : 27/04/2023 19:46:15
BlobName    : v7-4-0-daily20230427-1
ReleaseTag  : v7.4.0-daily20230427.1
```

Figure 2.4 – Viewing release metadata

When PowerShell 7 starts, it checks the metadata, and if a new version is available (from GitHub), you will see a message something like this:

```
PowerShell 7.2.11
Copyright (c) Microsoft Corporation.

https://aka.ms/powershell
Type 'help' to get help.

    A new PowerShell stable release is available: v7.3.4
    Upgrade now, or check out the release page at:
        https://aka.ms/PowerShell-Release?tag=v7.3.4
```

Figure 2.5 – Checking metadata at startup

Microsoft Update (MU) can also update PowerShell 7 when a new version is available. At the time of writing, MU can service your version of PowerShell within two weeks of it being available on GitHub. Likewise, if your installation is via the Microsoft Store, the Store should automatically download and apply updates.

These automatic methods of updating PowerShell 7 do take time, so if up-to-dateness is a key issue, consider just using the `Install-PowerShell.ps1` script when and where you need to update the version of PowerShell on your system immediately.

If you use other tools, such as Chocolatey, you have to use these tools when PowerShell notifies you that an updated version is available.

The three key pillars of PowerShell

PowerShell has three key pillars:

- **Cmdlets**: Small programs that do useful things, such as retrieve a set of files in a folder. Some cmdlets come with PowerShell, some come with applications and services, and you can leverage a huge library of third-party tools.

- **Objects**: Data structures representing entities within your computer and containing properties and methods. Cmdlets can consume and produce objects.

- **The pipeline**: The pipeline enables you to chain two cmdlets – the output of one cmdlet is sent, or piped, to a second cmdlet.

Cmdlets

Cmdlets are small programs that do useful things, such as getting the details of all the running processes. Cmdlets developers write these cmdlets as .NET classes, typically using C#.

Cmdlets come either with PowerShell itself or as part of an application such as VMware or the various Windows Server features. In *Chapter 5*, you can read more about the tools you can use to manage Windows, including the **Remote Server Administration Tools** (**RSAT**).

Cmdlets are named using a strict noun-verb syntax, based on a restricted and well-known set of verbs. For example, you use the `Get-Process` command to get details of the processes. Likewise, you would use the `Get-Service` command to get details of all the services on a system. The strict naming of cmdlets is a great feature that helps you to discover other cmdlets.

Cmdlets take parameters that affect how the cmdlet operates. You specify a parameter with a parameter name (which always begins with a - character) and usually some value. For example, if you wanted to get details on the DHCP client service running in Windows 11, you would type as follows:

```
PS C:\foo> Get-Service -Name Dhcp

Status    Name                 DisplayName
------    ----                 -----------
Running   Dhcp                 DHCP Client
```

Figure 2.6 – Using Get-Service to view a Windows service

For more details on Powershell cmdlets, see `https://packt.link/f9ZTD`.

Objects

In PowerShell, an object is a data structure that contains properties and methods about some entity, such as a file or a Windows process. The properties of an object are specific attributes of that object, such as the file's full name or the process's current CPU usage. You can create objects using cmdlets (for example, the `Get-Process` command returns objects of the `system.process.diagnostics` type).

You use objects in PowerShell when you manage Windows and write scripts to automate some activity, such as deleting all the files in temporary folders. Objects are fundamental to PowerShell and are great at simplifying scripting.

A benefit of objects is that the details of the object are easy to view. Just pipe the output of a cmdlet to `Get-Member`, and you can discover precisely what is inside each object. There is no *prayer-based text parsing*, as is more usual in Linux environments. See `https://packt.link/KULU6` for an explanation of prayer-based parsing.

For example, you can get details of the optional features available In Windows 11 using the `Get-WindowsOptionalFeature` cmdlet. When you use this cmdlet, PowerShell returns an array of objects, each representing one of the optional features. You can then pipe the output of that command to `Get-Member` to show what is inside each object occurrence like this:

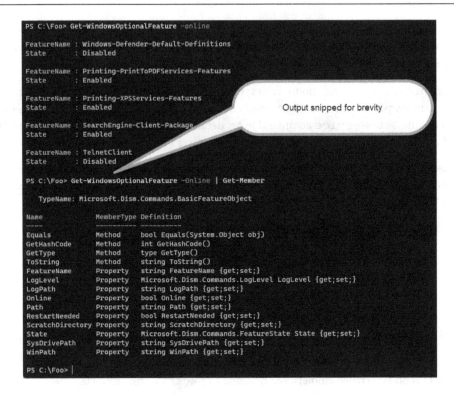

Figure 2.7 – Using Get-Member

When you automate Windows optional feature management, you easily discover that the property's name, holding the feature's current status is state. As you use PowerShell, this behavior becomes more and more useful.

For more details about objects inside PowerShell, see https://packt.link/QKxyh.

The pipeline

The pipeline is a feature of PowerShell that takes the objects a command creates and uses them as input for another PowerShell command. You use the | character to indicate the pipe operation, which you saw previously when you piped the output of the Get-WindowsOptionalFeature to the Get-Member command. The first cmdlet produced several objects (one for each Windows optional feature). By sending those objects to the next cmdlet, Get-Member can tell you what those objects look like.

The incredibly powerful pipeline enables you to create simple scripts to accomplish complex tasks. For example, suppose you wish to know which company made the software that uses the most virtual memory on your system. On Windows 11, each running application uses one or more (or a lot more) Windows processes. So, we can do this:

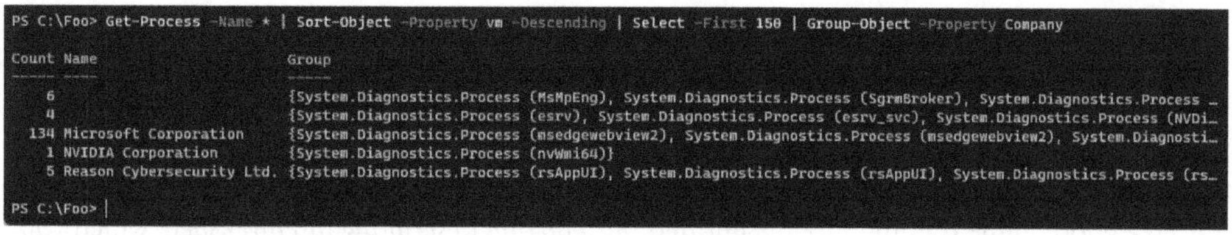

```
PS C:\Foo> Get-Process -Name * | Sort-Object -Property vm -Descending | Select -First 150 | Group-Object -Property Company

Count Name                  Group
----- ----                  -----
    6                       {System.Diagnostics.Process (MsMpEng), System.Diagnostics.Process (SgrmBroker), System.Diagnostics.Process …
    4                       {System.Diagnostics.Process (esrv), System.Diagnostics.Process (esrv_svc), System.Diagnostics.Process (NVDi…
  134 Microsoft Corporation {System.Diagnostics.Process (msedgewebview2), System.Diagnostics.Process (msedgewebview2), System.Diagnosti…
    1 NVIDIA Corporation    {System.Diagnostics.Process (nvWmi64)}
    5 Reason Cybersecurity Ltd. {System.Diagnostics.Process (rsAppUI), System.Diagnostics.Process (rsAppUI), System.Diagnostics.Process (rs…

PS C:\Foo>
```

Figure 2.8 – Using the pipeline

In this example, you use Get-Process to get all the processes on your system. Powershell returns a process object for each Windows process. You then pipe it to Sort-Object to sort the objects based on VM usage (with the greatest VM usage sorted to the top). Then you take the top 150 of those processes (that is, the 150 processes using the most VM) and group them by the company attribute of the process object, which should be the application manufacturer. However, some apps do not populate that property!

PowerShell rests on top of .NET, so each PowerShell object, each service, each process, and so on is a .NET object. .NET provides a rich set of objects that enable you to interact with all the key Windows services and applications. In many cases, PowerShell is merely a wrapper around the functionality provided within .NET.

For more information on the pipeline, see https://packt.link/QVl4S.

Understanding these three pillars is fundamental to learning and mastering PowerShell.

PowerShell's scripting language

PowerShell has a powerful scripting language that you can utilize to script key administrative tasks. The language is simple, with the syntax largely coming from C#.

The scripting language contains a variety of constructs that enable you to create powerful scripts. There are several features of the language:

- **Variables**: You use these to store data values in your scripts or console session. For more details on variables in PowerShell, see https://packt.link/qD2rO.

  PowerShell comes with several built-in variables, some of which control preferences within PowerShell. For details, see https://packt.link/iuvBC.

- **Operators**: PowerShell provides a rich set of arithmetic, comparison, and string operators. PowerShell 7 has implemented many additional operators unavailable in Windows PowerShell. For more details on PowerShell operators, see https://packt.link/oORPF.

- **Loops**: There are various ways to loop, iterating over a collection of objects to perform some action on a set of objects, such as changing the office name for a certain group of AD users. For details on loops in Powershell, see https://packt.link/d36fF and https://packt.link/zNi8F.

- **Flow of control**: Like all programming languages, PowerShell implements several different flows of control mechanisms. These allow you to alter the flow of control should some condition exist. For more details, see `https://packt.link/9ghy2`.

This chapter does not explore all aspects of the PowerShell scripting language fully. Microsoft has produced a good online training module entitled *Introduction to scripting in PowerShell*, which you can find at `https://packt.link/8hhgt`.

PowerShell's formatting features

PowerShell provides a wealth of formatting features, which are very useful in automation scenarios. You can format a string to include, for example, a user name, date/time, or some other value. You have complete control over how PowerShell should format the string.

PowerShell also formats objects for easy console output. When you send objects to the console, PowerShell uses built-in defaults, describing how to output any given .NET/PowerShell object. You can change those defaults, should you need to.

Like most programming languages, PowerShell supports a range of mechanisms to format a string. You can concatenate strings or insert the value of a variable into a string. For example, you can do this:

```
PS C:\Foo> $String1 = 'Jerry'
PS C:\Foo> $String2 = 'Garcia'
PS C:\Foo> "Hello $String1 $String2"
Hello Jerry Garcia
```

Figure 2.9 – String formatting in PowerShell

.NET Framework and .NET provide rich string formatting features, which are useful when you need precise control over the formatting. You can read more about PowerShell and string formatting at `https://packt.link/fJroH`.

Getting help

One feature that stood out when Microsoft first unveiled Windows PowerShell was the built-in help system. PowerShell comes with a `Get-Help` cmdlet. If you run it with no parameters, you can see a description of how to use the `Get-Help` cmdlet. But if you use `Get-Help` with the name of a cmdlet, you get help with that cmdlet. The built-in help system is invaluable, as it tells you what the cmdlet does, which inputs it takes, and what it outputs. With thousands of cmdlets to choose from, using `Get-Help` is much simpler than trying to remember them all, as shown in the following screenshot:

```
PS C:\Foo> get-help get-process

NAME
    Get-Process


    Get-Process [[-Name] <String[]>] -IncludeUserName [<CommonParameters>]

    Get-Process -IncludeUserName -InputObject <Process[]> [<CommonParameters>]


DESCRIPTION
    The `Get-Process` cmdlet gets the processes on a local computer.

    Without parameters, this cmdlet gets all of the processes on the local computer. You can also specify a particular process
    by process name or process ID (PID) or pass a process object through the pipeline to this cmdlet.

    By default, this cmdlet returns a process object that has detailed information about the process and supports methods that
    let you start and stop the process. You can also use the parameters of the `Get-Process` cmdlet to get file version
    information for the program that runs in the process and to get the modules that the process loaded.


RELATED LINKS
    Online Version: https://docs.microsoft.com/powershell/module/microsoft.powershell.management/get-process?view=powershell-7&W
    T.mc_id=ps-gethelp
    Debug-Process
    Get-Process
    Start-Process
    Stop-Process
    Wait-Process

REMARKS
    To see the examples, type: "Get-Help Get-Process -Examples"
    For more information, type: "Get-Help Get-Process -Detailed"
    For technical information, type: "Get-Help Get-Process -Full"
    For online help, type: "Get-Help Get-Process -Online"
```

Figure 2.10 – Using Get-Help

The help information quickly references what the cmdlet does and its calling syntax. You can also get examples and more detailed information, as shown in *Figure 2.10*.

You can also add help information to your scripts, which enables a script's user to use Get-Help and get details about the script. You achieve this by adding a special block of help text at the start of the script.

Another great feature of PowerShell's help system is that you can update the help text. The PowerShell help system enables the authors of the various scripts and commands you use to update help text and place it online. Then you use the Update-Help cmdlet to download the updates onto your system.

You may also note that there is very little help text the first time you use PowerShell. The reason for this is that PowerShell, by default, ships with minimal help text, which reduces the size of the PowerShell installation package. Once you first use any version of PowerShell on a system, you can download the most up-to-date help information, but only where you need it.

For more information on PowerShell's help system, see https://packt.link/xL902.

Modules and commands

In PowerShell, you use commands to have PowerShell carry out some operation, whether it be displaying a file, adding a user to Active Directory, or setting the IP address of a computer.

Commands

In PowerShell, a command is something you can run from a script file or at the console that performs some operation. There are several types of PowerShell commands:

- **Cmdlets**: These are small programs (technically written as a .NET class).
- **Functions**: These are, in effect, script cmdlets, which are small scripts that emulate the behavior of a cmdlet. You can think of them as script cmdlets.
- **Scripts**: These are sets of PowerShell commands, which you save with the extension .ps1.

In addition, the term command also applies to the many Win32 console applications, shipped with Windows 11 (over 300). Many of these are familiar to IT professionals, such as ipconfig.exe and ping.exe).

All PowerShell commands can accept and produce objects. The Win32 console applications, on the other hand, only take parameters (which tend to be unique to each console application) and output a string.

Console applications are very useful, although they can be harder to use in an automation scenario. If you need to, you can capture the output of a console application and use string manipulation to pull out the specific information you need. This approach, often called *prayer-based parsing*, is common in the Unix/Linux world. PowerShell objects make this process easier, and in most cases, there are PowerShell commands that can replace old console applications.

Modules

In PowerShell, a module contains a set of PowerShell commands, module metadata, and other information (such as help or formatting information). Before using any PowerShell command, PowerShell first loads the module containing that command.

By default, if you use a command in a module stored in one of a set of well-known folders, PowerShell automatically loads the module and then executes the command. Powershell has several default locations from which to load modules. The Windows PSModulePath environment variable defines PowerShell's paths to discover and load modules (by default). Every time PowerShell starts up, it examines the current value of the module path variable and determines all the commands available (in those locations). This process creates a cache of all commands and which module to use.

For example, to use the Get-Process command to view all the running processes, PowerShell needs to load the Microsoft.PowerShell.Management module. This is the module that contains this command. Once PowerShell loads the module, it can execute the command.

Each time PowerShell starts up, it caches the location of every command in every module (for those modules located in one of those well-known places). This means that, in most cases, you only need to type a command, and the rest of the magic happens behind the scenes. You can put a module into any location on the disk and manually load it using `Import-Module`.

You can get modules from a variety of places:

- **Built-in modules**: These are modules that come with Windows PowerShell and PowerShell 7. They provide the basic PowerShell functionality.

- **With applications or features**: Some applications or Windows optional features come with PowerShell management tools. These enable you to manage the feature or the application. See *Chapter 5* for more details on the RSAT.

- **From the PowerShell Gallery**: This is a Microsoft-maintained repository of community-authored modules. There are thousands of modules to choose from, as you can see at `https://packt.link/ghY17`.

- **Commercial software offerings**: Some software firms provide commercial modules. For example, /n software offers a suite of cmdlets with great communication capabilities. For more information about the cmdlets, see `https://packt.link/FLF11`.

- **DIY**: You are free to develop and use modules you have written or that your organization has developed. DIY modules exist (in the PowerShell Gallery) for a huge range of uses. For example, a Grateful Dead enthusiast created a module of scripts (`https://packt.link/gq1Lb`). These scripts may not help you manage Windows 11, but they indicate the breadth of available modules. Seek, and ye shall find, as the saying goes.

No matter where you get them from, modules are a critical component of PowerShell that help the authors of PowerShell package and ship their PowerShell modules.

If you want to learn more about building your own PowerShell 7 module, see `https://packt.link/wvznY`.

Discovery

In PowerShell, the term *discovery* means using built-in commands to find information on how to use PowerShell and PowerShell commands. The more you can discover, the less you need to memorize. Discovery is an extremely useful feature that the PowerShell team designed in the first versions and refined later. Without discovery, using thousands of commands and hundreds of potential modules would be much more difficult.

There are four key aspects of discovery within PowerShell:

- **Discovering modules**: Since modules contain commands, finding likely modules is useful. You can use the `Get-Module -ListAvailable` cmdlet to see all the modules that PowerShell can discover on your host. You can also use `Find-Module` to find the modules contained in the PowerShell Gallery that may be interesting.

- **Discovering commands**: Once you have a module, you can find the commands within the module by using Get-Command and specifying the module name like this:

```
PS C:\Foo> Get-Command -Module BitsTransfer

CommandType     Name                           Version     Source

Cmdlet          Add-BitsFile                   2.0.0.0     BitsTransfer
Cmdlet          Complete-BitsTransfer          2.0.0.0     BitsTransfer
Cmdlet          Get-BitsTransfer               2.0.0.0     BitsTransfer
Cmdlet          Remove-BitsTransfer            2.0.0.0     BitsTransfer
Cmdlet          Resume-BitsTransfer            2.0.0.0     BitsTransfer
Cmdlet          Set-BitsTransfer               2.0.0.0     BitsTransfer
Cmdlet          Start-BitsTransfer             2.0.0.0     BitsTransfer
Cmdlet          Suspend-BitsTransfer           2.0.0.0     BitsTransfer
```

Figure 2.11 – Discovering commands in a module

- **Discovering cmdlet details**: Once you know the commands, you can learn how a given command works – what parameters you can specify, what it outputs, and most importantly, what this command does. Using Get-Help, as noted earlier, helps you to work out what a cmdlet, function, or script does. For example, you can determine more details of the Start-BitsTransfer command like this:

```
PS C:\Foo> Get-Help -Name Start-BitsTransfer

NAME
    Start-BitsTransfer

SYNOPSIS

DESCRIPTION
    The Start-BitsTransfer cmdlet creates a Background Intelligent Transfer Service (BITS) transfer job to transfer one or more
    files between a client computer and a server. The TransferType parameter specifies the direction of the transfer. By
    default, after the cmdlet begins the transfer, the command prompt is not available until the transfer is complete or until
    the transfer enters an error state. If the state of the returned BitsJob object is Error, the error code and description
    are contained in the object and can be used for analysis.

    The Start-BitsTransfer cmdlet supports the download of multiple files from a server to a client computer, but it does not
    generally support the upload of multiple files from a client computer to a server. If you need to upload more than one
    file, you can use the Import-Csv cmdlet to pipe the output to the Add-BitsFile cmdlet to upload multiple files. Or, if you
    need to upload more than one file, consider a cabinet file (.cab) or a compressed file (.zip).

RELATED LINKS
    Online Version:
    Add-BitsFile
    Complete-BitsTransfer
    Get-BitsTransfer
    Remove-BitsTransfer
    Resume-BitsTransfer
    Set-BitsTransfer
    Suspend-BitsTransfer

REMARKS
    To see the examples, type: "Get-Help Start-BitsTransfer -Examples"
    For more information, type: "Get-Help Start-BitsTransfer -Detailed"
    For technical information, type: "Get-Help Start-BitsTransfer -Full"
    For online help, type: "Get-Help Start-BitsTransfer -Online"
```

Figure 2.12 – Getting help on a cmdlet

The help text gives you more details about what a command does.

- **Discovering what a cmdlet returns (that is, the objects emitted by the command)**: If you have a cmdlet that returns a value, remember that what you see in the console represents the objects returned from the cmdlet. The actual objects may contain more information that PowerShell does not display by default. To discover the details of the objects returned, you can do this: if you are using Windows 11's Hyper-V function, you can use Get-VM to determine the VMs on your host.

```
PS C:\Foo> Get-VM | Get-Member

   TypeName: Microsoft.HyperV.PowerShell.VirtualMachine

Name                                    MemberType   Definition
----                                    ----------   ----------
CheckpointFileLocation                  AliasProperty CheckpointFileLocation = SnapshotFileLocation
ParentCheckpointId                      AliasProperty ParentCheckpointId = ParentSnapshotId
ParentCheckpointName                    AliasProperty ParentCheckpointName = ParentSnapshotName
VMId                                    AliasProperty VMId = Id
VMName                                  AliasProperty VMName = Name
Equals                                  Method       bool Equals(System.Object obj)
GetHashCode                             Method       int GetHashCode()
GetType                                 Method       type GetType()
ToString                                Method       string ToString()
AutomaticCheckpointsEnabled             Property     bool AutomaticCheckpointsEnabled {get;}
AutomaticCriticalErrorAction            Property     Microsoft.HyperV.PowerShell.CriticalErrorAction AutomaticCriticalErrorAction {get;}
AutomaticCriticalErrorActionTimeout     Property     int AutomaticCriticalErrorActionTimeout {get;}
AutomaticStartAction                    Property     Microsoft.HyperV.PowerShell.StartAction AutomaticStartAction {get;}
AutomaticStartDelay                     Property     int AutomaticStartDelay {get;}
... snipped for brevity
```

Figure 2.13 – Getting the members of a class

These four discovery approaches help you use PowerShell – to find modules and commands, determine what a PowerShell command does, and establish which outputs, if any, result.

PowerShell and security

PowerShell security is an issue for most IT professionals. PowerShell comes with various security features that can help you both harden PowerShell and improve your environment's security.

Security by default

The PowerShell team designed PowerShell to be secure by default. But it provides IT professionals with great power at the same time. And with great power comes great responsibility.

Windows has a wealth of access control mechanisms, such as the **Access Control List** (**ACL**) with **Windows NT File System**. As an administrator, setting the ACL on files and objects in Active Directory is an important line of defense. If you set an ACL that denies some user access, that user cannot use PowerShell to bypass the ACL. Likewise, numerous commands require that you run PowerShell as an administrator.

But anyone possessing administrator credentials can do almost anything on a system.

PowerShell logging

PowerShell records a variety of actions in Windows event logs automatically. These include the PowerShell engine starting up, shutting down, and more. Windows PowerShell and PowerShell 7 log the same events but use different Windows event logs.

Windows PowerShell sends event details to the event log, **Windows PowerShell** (underneath **Applications and Services Logs** in **Event Viewer**). PowerShell 7 writes event log entries to the **PowerShellCore** event log (also underneath **Applications and Services Logs** in **Event Viewer**).

From a forensic point of view, these event log entries may be useful to discover when someone uses PowerShell on a given host. To learn more about PowerShell and logging, see `https://packt.link/CpBO7`.

PowerShell script block logging

In addition to logging basic PowerShell engine events, PowerShell can also log any script block executed on a given host. This form of logging captures a script's complete activity and content, providing a full audit trail of a PowerShell session.

Powershell does not enable logging by default since a performance overhead is associated with event logging, particularly script block logging. You can change this manually on a given host or via Group Policy.

The **about_Logging** page, noted previously, contains more information about script block logging. You can also refer to the following blog post: `https://packt.link/hq0tE`.

Module logging

You can turn on logging for specified PowerShell modules. If you turn on module logging for a given module, PowerShell logs pipeline execution events into the event log.

Execution policy

PowerShell has an execution policy that controls a user's ability to run any PowerShell script directly. You can use the `Set-ExecutionPolicy` to change the policy and `Get-ExecutionPolicy` to view the current policy.

You can set the execution policy to different values:

- `AllSigned`: All scripts or configuration files must be signed by a trusted publisher.
- `RemoteSigned`: Scripts and configuration files downloaded by the internet must be signed by a trusted publisher, while you can execute scripts authored locally.

- `Restricted`: PowerShell does not load configuration files or run any scripts. Note: This is the default for Windows 11 hosts!

- `Unrestricted`: PowerShell runs any script and loads all configuration files.

- `Bypass`: PowerShell blocks nothing and issues no warning messages.

- `Default`: Sets the execution policy to **Restricted** on Windows 11 hosts (and sets the policy to **RemoteSigned** on Windows Server systems).

Note that the execution policy is *not* a security barrier, but more of a minor speed bump. While a restrictive policy may stop you from running a script, you can open the file in a text editor such as Notepad, copy all the text, paste it into PowerShell, and PowerShell runs the script. For a more detailed look at execution policies, see the help file at `https://packt.link/JLOd6`.

Many (and possibly most) administrators use `Set-ExecutionPolicy` to set the policy on the local machine to `Unrestricted` and perform the administrative task. Sometimes, they may even remember to set it back to the default. As noted in the help document, you can also use the `-Scope` parameter to limit the change of the execution policy in line with the current process.

In large production environments, signing the scripts you use in the enterprise, using your firm's private key infrastructure is a best practice. You, or your security team, would need to create an X.509 certificate that enables code signing. Then you can use the `Set-AuthenticodeSignature` to sign scripts.

If you plan to employ code-signing PowerShell scripts, you must clearly define the process of script signing, including detailed testing and vetting. You must ensure a good script signing process, including who can access the signing key (and how). If the certificate is put on a local file server for anyone to use, code signing achieves nothing.

Transcription

You can use the `Start-Transcript` cmdlet to create a detailed plain text record of all or part of a PowerShell session. You can either set this in a script (to memorialize the execution of a script) or within a PowerShell session to record the details of the session. You use `Stop-Transcript` to stop the recording.

You can also enforce transcripts of all PowerShell sessions via Group Policy. If you do a lot of transcription, remember to implement a process to review the transcripts and clear or archive older files.

Configuring PowerShell

PowerShell comes with a default configuration. To further configure your environment and override that default configuration, you can add modules, use PowerShell profile scripts, and deploy PowerShell settings via Group Policy.

Adding modules

PowerShell and Windows applications come with numerous modules (each containing PowerShell commands). To extend PowerShell's capabilities, you can find and download additional modules. The PowerShell Gallery is home to thousands of community (and Microsoft) authored modules that could be helpful, as noted in the *Modules* section earlier.

You can use your browser to view the contents of the gallery at `https://packt.link/ghY17`. The gallery also includes newer versions of modules shipped with Windows (and Windows PowerShell). For example, the authors of both the `PackageManagement` and `PowerShellGet` modules have created updated versions that you can download.

The `NTFSSecurity` module, for example, makes it much easier to automate the ACLs within Windows NTFS. The commands in this module greatly simplify the management of ACLs on files and folders and the configuration of permission inheritance.

To use PowerShell with the PowerShell Gallery, you need the latest version of the `PowerShellGet` module. You can also get this update from the PowerShell Gallery.

Profiles

Profiles are PowerShell scripts that PowerShell runs automatically each time you run either Windows PowerShell or PowerShell 7. The profile files are in a well-known location and have a well-known name (well known to PowerShell, that is).

To cater to a wide variety of usage scenarios, PowerShell defines four separate profile files:

- `AllUsersAllHosts`: PowerShell runs this profile file for every user using any PowerShell host (including Windows Terminal, VS Code, etc.).
- `AllUsersCurrentHost`: Powershell runs this script for all users using this specific PowerShell host.
- `CurrentUserAllHosts`: PowerShell runs this solely for the currently logged-on user for all hosts.
- `CurrentUserCurrentHost`: PowerShell runs this solely for the current logged-on user and only this host. The built-in `$Profile` variable holds the name of this profile file.

You can discover the location of each of these four files (including the full path name) by piping `$Profile` to `Format-List`, as follows:

```
PS C:\Foo> $Profile | Format-List -Force

AllUsersAllHosts      : C:\Program Files\PowerShell\7\profile.ps1
AllUsersCurrentHost   : C:\Program Files\PowerShell\7\Microsoft.PowerShell_profile.ps1
CurrentUserAllHosts   : C:\Users\tfl.COOKHAM\Documents\PowerShell\profile.ps1
CurrentUserCurrentHost : C:\Users\tfl.COOKHAM\Documents\PowerShell\Microsoft.PowerShell_profile.ps1
Length                : 74
```

Figure 2.14 – Profile file locations

For more details on the PowerShell profile, see `https://packt.link/3q3ek`.

Group Policy

Group Policy is a feature of Windows Active Directory. Group Policy enables the administrator to dictate specific settings and policies on individual systems. Windows automatically applies these policies each time you restart the system or log on and refresh the policies regularly. Group Policy is a great way to leverage Active Directory to create customized, desktop environments.

The PowerShell-related policies you can set using Group Policy are as follows:

- **Execution Policy**: Specifies a value different from PowerShell's default execution policy
- **Module Logging**: Whether to perform module logging and for which modules
- **Script Block Logging**: Whether to log script block execution
- **Transcription**: Whether to create transcripts for all PowerShell sessions
- **The default source path for Update-Help**: Allows you to create a local repository for help information and to have `Update-Help` use that location to obtain the updates

See `https://packt.link/x4M1W` for more information about these Group Policy settings.

Using PowerShell

Traditionally, with Windows PowerShell, you have two main ways of accessing PowerShell: the PowerShell console and the Windows PowerShell ISE. The ISE was popular and was designed to be extendable.

Developers could also host and access the PowerShell engine. This architecture allows an application to leverage PowerShell. For more details on hosting PowerShell, see `https://packt.link/CQAVf`.

The ISE, however, does not support PowerShell 7 and Microsoft has developed an outstanding replacement, the Microsoft Terminal. You can use this as a high-fidelity way to access Windows PowerShell and PowerShell 7.

The PowerShell console

In the early versions of PowerShell, the Windows PowerShell console was the only way to access it. Strictly speaking, the console is an application (`conhost.exe`) that exposes the Windows PowerShell engine to the end user and looks like this:

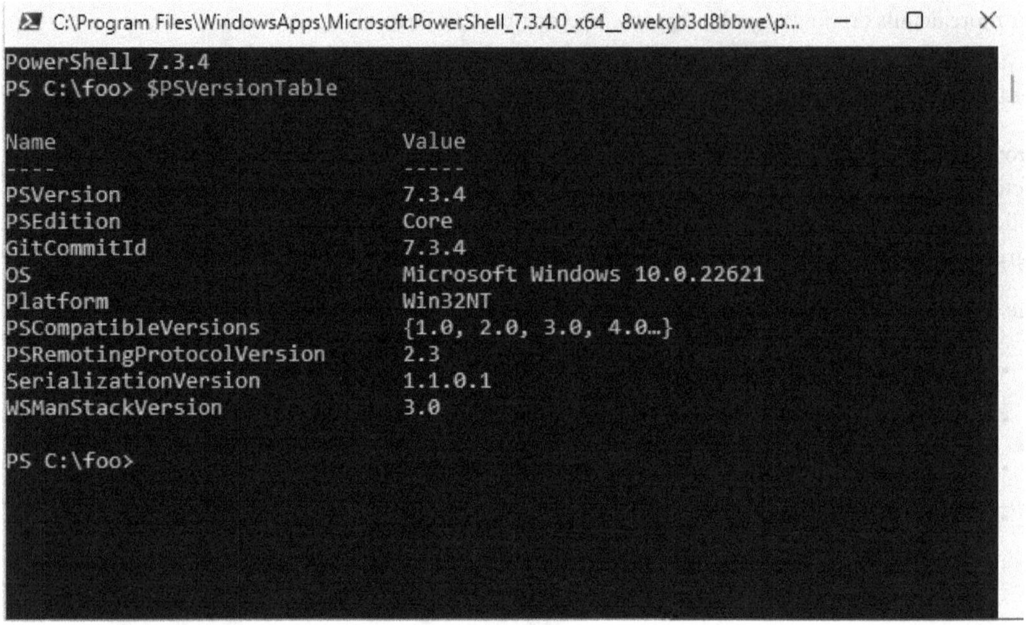

```
C:\Program Files\WindowsApps\Microsoft.PowerShell_7.3.4.0_x64__8wekyb3d8bbwe\p...    —    □    ✕
PowerShell 7.3.4
PS C:\foo> $PSVersionTable

Name                            Value
----                            -----
PSVersion                       7.3.4
PSEdition                       Core
GitCommitId                     7.3.4
OS                              Microsoft Windows 10.0.22621
Platform                        Win32NT
PSCompatibleVersions            {1.0, 2.0, 3.0, 4.0…}
PSRemotingProtocolVersion       2.3
SerializationVersion            1.1.0.1
WSManStackVersion               3.0

PS C:\foo>
```

Figure 2.15 – PowerShell console

The Windows PowerShell Integrated Scripting Environment (ISE)

Microsoft first introduced the ISE with Windows PowerShell V2 but made substantial improvements with Windows PowerShell V3. The ISE combines a good script editor with a console window. In Windows, this application is at C:\WINDOWS\System32\WindowsPowerShell\v1.0\powershell_ise.exe.

It's worth noting that while you can continue to use the ISE, Microsoft has ceased active development of the product. The replacement is Visual Studio Code.

The ISE is a good tool that enables you to create, test, and use PowerShell scripts. The ISE looks like this:

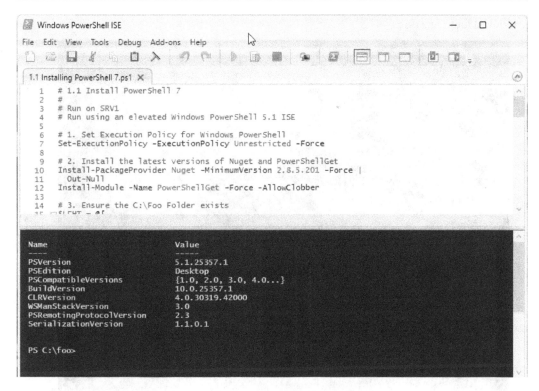

Figure 2.16 – The ISE

Windows Terminal

The original console application did not allow Microsoft to extend the console. As a replacement, Microsoft developed Windows Terminal. The goal was to create a great terminal, capable of hosting cmd.exe, Windows PowerShell, and PowerShell 7 (and more). The terminal also supports a tabbed interface.

The terminal looks like this:

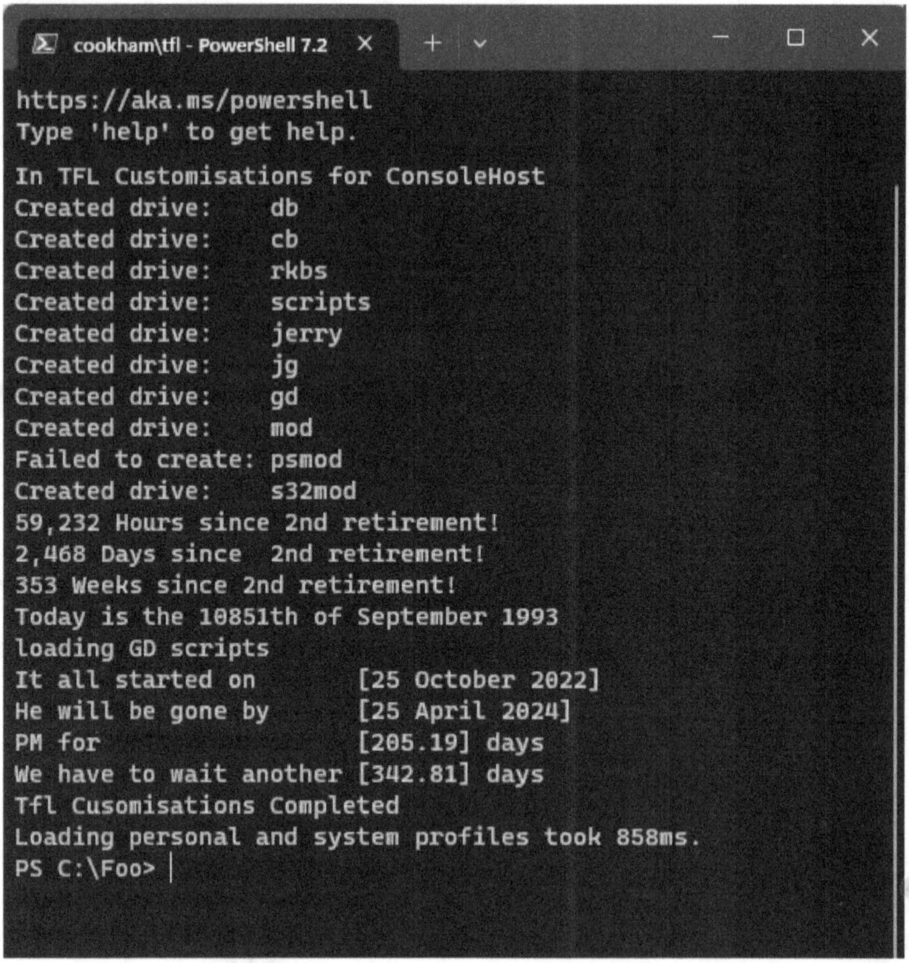

Figure 2.17 – Windows Terminal

Windows 11 also allows you to use the console or Windows Terminal. You access this by using the Windows 11 **Settings** apps and navigating to the **For developers** page, like this:

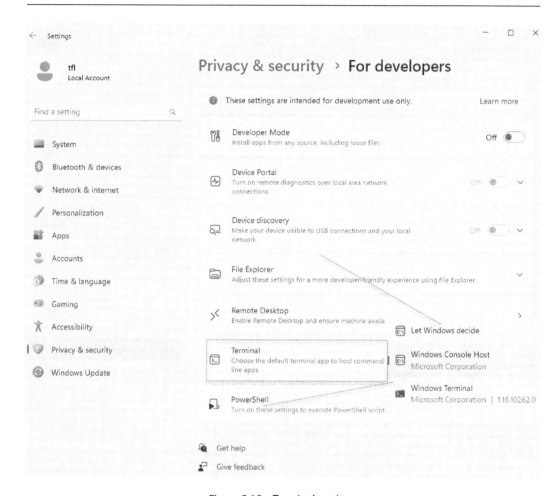

Figure 2.18 – Terminal settings

For a more detailed overview of Windows Terminal, see https://packt.link/BQi45.

Microsoft Visual Studio Code

As part of the move away from the Windows PowerShell ISE, Microsoft launched a new tool, Visual Studio Code. Visual Studio Code looks a bit like the ISE, with a three-pane layout (folders and files on the left, an editor pane in the top right, and the terminal in the bottom-right pane).

Visual Studio Code looks like this:

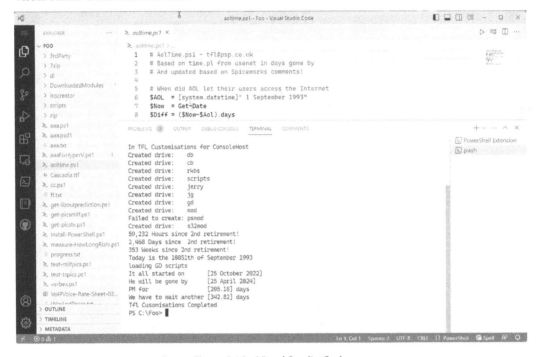

Figure 2.19 – Visual Studio Code

Visual Studio Code is not loaded into Windows 11 by default, but you can use the Windows Store to install the product. You can also download the latest versions directly from GitHub at https://packt.link/Y5CEr. Or you can use your favorite package manager to install Visual Studio Code.

For more information on Visual Studio Code or to download the product, see https://packt.link/QfvFZ. Once installed, Visual Studio Code detects available updates and prompts you to install them.

One great feature of Visual Studio Code is the extension architecture. This allows you to add a huge range of extensions, greatly enhancing your usage. For example, you can add a spell-checker, a great drawing package, Git Integration, and support for Markdown. For more details on the available extensions and how to manage Visual Studio Code extensions, see https://packt.link/K13Jv.

Desired State Configuration

Desired State Configuration (DSC) is a management framework in Windows PowerShell that provides a standardized way of defining a system's DSC. This enables the complete automation of device configuration using a declarative model: you create a configuration that states how a device should be configured, publish that configuration, and then wait for the devices to update themselves to match the configuration. DSC, as described here, is only available using Windows PowerShell.

A single configuration can be used across multiple devices, ensuring all hosts have identical and standardized settings. DSC prevents configuration drift when numerous changes occur over time.

DSC supports two deployment methods:

- **Push mode**: In this mode, the administrator makes the configurations and manually pushes them to the target devices. Push mode is one-way communication and can only work if the devices are available on the network during the push. This option best suits small environments where all devices are always connected. It also assumes that the appropriate PowerShell modules are available on all devices.

- **Pull mode**: In this mode, the administrator creates a pull server that hosts the configuration information. You then configure each device to contact this server at regular intervals (such as every 15 minutes) to look for any configuration changes. The device can pull the DSC data and any required modules if found. Using pull mode allows you to automate the deployment of configurations to a high frequency and doesn't require further interaction from the administrator.

DSC, in Windows PowerShell, has both a configuration and a resource side, and a configuration manager runs the show between the two:

- **Configurations**: These are PowerShell scripts that define and configure the resources. The DSC engine will examine the configuration to ensure the system adopts the desired state.

- **Resources**: A resource is a code that keeps the target of a configuration in a specific state. By default, DSC has several resources, which you can see by running the Get-DSCResource cmdlet. These are typically in Windows PowerShell modules:

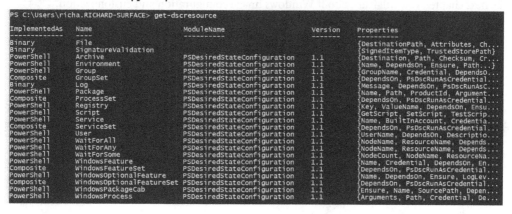

Figure 2.20 – Getting DSC resources

- **Local Configuration Manager** (**LCM**) runs on the local device to ensure the configuration is applied correctly. The LCM polls the system to ensure that the state defined by the configuration is maintained, and corrects any errors.

An example of how you would use DSC configurations would be to prevent a specific process from running or having a registry key or security policy set in a particular way to meet your security requirements.

Use the official Microsoft DSC resource kit for more advanced configuration: `https://packt.link/ZpjZw`.

At the time of writing, DSC does not provide the rich features available with Windows PowerShell. Beginning with PowerShell V 7.2, PowerShell no longer ships with the `PSDesiredStateConfiguration` module as part of the PowerShell 7 download. This has allowed a separate DSC team to invest in and develop DSC independently of PowerShell (and has the benefit of reducing the size of the overall PowerShell download package).

While you can do some DSC-based configuration, DSC for PowerShell 7 does not currently include the richness available in Windows PowerShell. For more details on DSC and Powershell, see `https://packt.link/CBsp0`.

Summary

To summarize this chapter, we have learned about Windows PowerShell and PowerShell. We examined the components of PowerShell, the PowerShell scripting language, and modules and commands. We then learned how to manage security with PowerShell. We also looked at configuring and using PowerShell. Finally, the chapter ended with a brief look at DSC. There is a lot more to PowerShell, such as classes, error handling, debugging, and more. For a more detailed look at PowerShell, see `https://packt.link/SNGDk`.

In the next chapter, you will learn about Windows image configuration and customization.

3

Configuration and Customization

In this chapter, you'll learn best practices for image creation and customization. The methods discussed will primarily be applicable to the Windows 11 Enterprise and Education editions. If your environment also includes the Professional edition, you will find that some recommended settings do not work or apply as expected. Microsoft used to maintain an index of settings that only apply to Windows 10 Enterprise and Education editions; with the release of Windows 11, guidance has been removed on these settings for some reason. Windows 11 Education Edition (or SE) does have a few settings that can be changed on the device, listed here: `https://packt.link/1BDUI`.

In this chapter, we will cover the following topics:

- Evolution of **Windows as a Service (WaaS)**
- Image customization
- Microsoft Autopilot
- Upgrade expectations
- Security mitigation
- **Security Compliance Toolkit (SCT)**
- Microsoft telemetry

Evolution of WaaS

Microsoft has continued to evolve the WaaS journey with the release of Windows 11. Considering this, to help enterprise environments keep up, Microsoft has made significant investments in tools and process development focused on deployment, Autopilot being probably the best example.

While the tools changed, the idea of configuring and tweaking an image without having to go through time-consuming task sequence steps and rigorous and methodical tweaking of settings is certainly a boon for the enterprise administrator (and perhaps a bane for the deployment-focused IT professional). It's my firm belief that IT is heading down a path where imaging (as we now know it) will be a thing of the past, where a Windows 11 machine can be plugged into a network, joined to Active Directory or **Azure Active Directory** (**Azure AD**), and policies are pushed down to configure the **user experience** (**UX**). I further suspect that eventually, a container-like technology will take hold where the user profile is just a container to load at login. Given the preponderance of badly applied folder redirection and roaming profile Group Policies in enterprise environments, this is probably a good thing, as many administrators contend with difficult or conflicting guidance on deployment, or even outsource their imaging work due to the complex nature of the work.

One of the aspects of WaaS that might not be anticipated by a lot of IT professionals (yet) is that things are going to change under the hood from build to build of Windows, more than likely. I'm not simply describing a UX design change or stability or anything such as that. I am speaking about the core of what Windows has been for enterprises for some time now.

Typically, enterprises are used to modifying the operating system to suit their needs. Need software to run an ATM? Great—Windows Embedded was always the answer. Want to launch missiles, view medical images, process payrolls, or any of the other myriad tasks workers in businesses, government organizations, and even homes do? Great—Microsoft Windows is for you!

What I am saying is that the deep customization knobs we are used to from the Windows XP and Windows 7 days are in some ways gone. That is not to say that edge cases are no longer welcome in Windows land—not at all. What it is saying is that for corner cases, rather than forcing the square peg into the round hole, they may be better served by using the appropriate tool available. For example, if you need Windows to be a kiosk, the use of assigned access is far preferred over hacking the registry in ways that may cause unintended issues later.

Microsoft is actively taking feedback on the changes it makes from build to build and modifying its roadmaps as a result of that feedback. Edge cases will still have a home in Windows, but it may be a different home than the old one.

So, while user profile customization seems to be headed down a new and exciting path, image customization is still available and can be necessary and worth the effort. The tools for this are the **Microsoft Deployment Toolkit** (**MDT**) and **Windows System Image Manager** (**Windows SIM**) from the **Windows Assessment and Deployment Kit** (**Windows ADK**).

The official statement from Microsoft is that the MDT is not a supported deployment method (see `https://packt.link/hqetr` for more information); however, it appears most folks that use MDT are happily ignoring this statement and instead using MDT to create images and deploy Windows 11.

Information on doing this is located at `https://packt.link/Wexry` and works fine. Expect to see this updated for Windows 11 23H2 in the coming weeks (from the date of writing).

Image customization

In an enterprise environment with many legacy applications and department configurations, deploying an image preconfigured and set up for the user makes a lot of sense. Standing up an MDT environment in an enterprise is a relatively easy task (usually, it takes more change in control/security procedures than actual install/setup time) that can be completed in an afternoon in most cases. Customizing the image is best done with reproducible tooling, and MDT will help with that as you can modify the default user profile.

This can be done for branding purposes or application delivery that stores itself entirely within the user profile folder structure. Once the customizations are completed, the profile is captured as the default administrator profile, so future logons inherit the customization.

Imaging process

Once your image is baked, you can take it and deploy it with **System Center Configuration Manager (SCCM)** or MDT or even give it to an **original equipment manufacturer (OEM)** to have placed on your computers purchased from them before you receive them. The process for baking an image is generally this:

- An environment is created that is off the production network. This is usually a virtualized environment and can even be all on a single host. A standalone **Dynamic Host Configuration Protocol (DHCP)** and artificial subnet with a **network address translation (NAT)** rule for the MDT host are preferred.

- A **virtual machine (VM)** is created that hosts the MDT server; 8 GB of RAM (recommended; 4 GB will work but performance may suffer) and a few processors are typically sufficient for image-creation purposes. A server OS is preferred for MDT but it can run on a client OS in a pinch.

- Another VM is created that will be your reference image container. It should be set up with 4 GB RAM and 2 processors, which is generally sufficient, though the process will likely go faster with 8 GB of RAM and 4 processors. This machine just needs to connect to the **Windows Server Update Services (WSUS)** and MDT hosts and mount an ISO produced from the MDT server process.

- MDT is used to build a reference image from the ISO of Windows 11 Enterprise, and a boot ISO is used to boot the VM reference container and run the task sequence to capture the completed **Windows image (WIM)** for later deployment.

Later deployment can be through any generally available deployment mechanism, such as MDT, Tanium, and SCCM (via **Operating System Deployment**, or **OSD**). **Windows Deployment Services** (**WDS**) has been partially deprecated. Microsoft made this announcement (`https://packt. link/EC2fr`):

> *The operating system deployment functionality of Windows Deployment Services (WDS) is being partially deprecated. Starting with Windows 11, workflows that rely on* `boot.wim` *from installation media or on running Windows Setup in WDS mode will no longer be supported.*

There are some considerations to this process that need to be reviewed, as follows:

- How often are you going to patch/capture your image? If you don't, eventually the image will be in a state where it deploys to hardware and then runs Windows updates for over 30 minutes before the system is usable for the end user. Generally, organizations create images for systems to speed deployment, and if you don't service the golden image with frequent updates, you'll end up not meeting your original goal.

- Are you going to do Zero-Touch or Lite-Touch deployment?

  - Zero-Touch is done via SCCM OSD or a third-party product and involves (usually) MAC address reservation for a specific image, or perhaps a user runs through a script that determines the appropriate image to lay down on hardware.

  - Lite-Touch is done when some prodding is needed to spur the deployment on. It can technically be fully automated by scripting some solutions and it is achieved with SCCM OSD/MDT/WDS or any other third-party tools available commercially.

These considerations should help guide you in making a robust imaging system that you can use to create and maintain your golden images.

Customizing the image

Customization in Windows 11 can be a mix of PowerShell scripts, Group Policies/Group Policy preferences, and registry key tweaks. The site `https://packt.link/k9JZa` is a boon for image customization, showing via a filter all the **Group Policy Objects** (**GPOs**) that can be tweaked on Windows 11, including which registry key is being set by the policy:

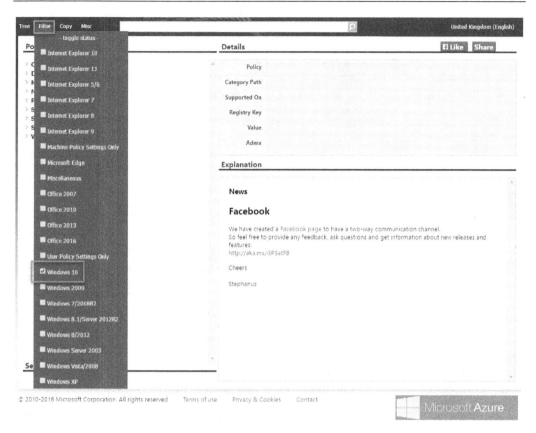

Figure 3.1 – Group Policy search

It is important to note with any customization effort that eventually, the administrator will run into a setting that cannot be edited or tweaked for all users by default. The Windows product group has determined that some settings are not for enterprises or admins to tweak but are instead user-only settings that are part of their personalization efforts.

But what if Group Policy or **Group Policy Preferences (GPP)** cannot be used to achieve your desired outcome? **Process Monitor (ProcMon)** logging while you configure the UI as desired may give a hint of a registry key that you need to modify (which can usually be added by Group Policy or REG ADD commands) to make the tweak happen. This is a frowned-upon practice, however, primarily because those registry keys may change location (or just not work) on a build-to-build basis of Windows 11. So, if you use GPOs to handle tweaking, you are on a much more solid, supported path for your image and UX than hacking the registry for undocumented setting tweaks.

Microsoft Autopilot

Microsoft Autopilot fills a gap in **Bring Your Own Device (BYOD)** scenarios, making IT organizations that use it more flexible on hardware acquisitions. Need to programmatically upgrade from the OEM-installed Windows Home SKU to Windows 11 Pro or Enterprise? No problem!

Microsoft Autopilot works by leveraging features available in Windows client OSes, Azure AD, and **mobile device management** (MDM) services such as Microsoft Intune, formerly **Microsoft Endpoint Manager** (MEM). Naturally, for someone to autopilot a computer from a typical store, an internet connection is required. DNS resolution for external names and ports 80 (HTTP), 443 (HTTPS), and 123 (UDP/NTP) need to be open.

The system in question authenticates with Azure AD. Once authentication happens, Microsoft Intune or another MDM solution can be used to trigger the enrollment of the device. Windows Update and Delivery Optimization are used to deliver updates, updated Microsoft Store applications, Office updates, Intune Win32 and MSIX apps, and **line-of-business** (LOB) applications.

Note that hybrid Azure AD domain join can be used, but only if the device is located on the internal hybrid network. Michael Niehaus does document a method to get around this, however: https://packt.link/u9D4F.

It's also worth noting that firmware TPM devices do not include all the needed certificates at boot and must retrieve updates from the manufacturer on the first device use. AMD, Intel, and Qualcomm all have hosts from which updates are automatically downloaded and applied.

Note that details are subject to change with each release of Windows and Autopilot, and it's best to consult the *Microsoft Learn* pages for details. Here are some links to get you started:

- Software requirements are located here: https://packt.link/8OWje.
- Network requirements can be found here: https://packt.link/15jTl.
- Licensing requirements are located here: https://packt.link/iLQLz.
- Configuration requirements are located here: https://packt.link/STP15.
- At the time of writing, self-deploying mode is in public preview. Details are here: https://packt.link/HSTPM.

BYOD scenarios

For BYOD scenarios, Microsoft Intune is the Microsoft-recommended vehicle for deployment. The suite will utilize integrated MDM policies to manage what happens to corporate data on a device when you determine the employee is no longer an employee, or if the device was stolen/is missing and you need to wipe it. Microsoft Intune is worth a book unto itself and is beyond the scope of this one. Just be aware that if BYOD is part of your endpoint strategy, you should be looking at Intune or a competing offering to manage this properly.

If you are put into a situation where you must implement BYOD without an MDM solution, be sure to consider the software licensing aspects of your implementation. Are you legally allowed to install the software on a machine that isn't actually yours? Do you really want to do that? It's interesting licensing and support boundary talk that needs to be ironed out, even with MDM. Not having a proper solution to manage it makes it very muddy indeed.

Upgrade expectations

Historically, when Windows upgraded, it carried all its baggage with it from the previous install (for better or for worse). Windows 10 and 11, however, seem to have deviated from this. Now, if an application is deemed incompatible with the build being upgraded to, the application will simply not be present in the post-upgrade operating system. One can use the `/Compat` command-line switch to automatically address compatibility issues. Documentation on this function is located here: `https://packt.link/fv3OZ`.

Windows should warn the user of this prior to upgrading and, if ignored, report this in a report file at `C:\Windows\Panther` named `miglog.xml` that the application was not migrated forward.

When first faced with this news, it is logical to assume that this is a complete disaster and a poor choice. However, consider the upgrade process as a guardian of sorts. The upgrade option will not be available if compatibility issues are present. If an application is going to break the installation, why migrate it? Also, note that I did not state the data would not be migrated. No—it is kept (if it is stored appropriately) in the user profile.

This makes the maneuvering of application compatibility between the OS and third-party software both problematic and somewhat pragmatic: problematic in that a slow software developer of a key enterprise application can demonstrably keep an upgrade from moving forward without significant application shims or other tricks, pragmatic because either the software works and therefore is migrated during the upgrade process or it doesn't work and won't be there to create a fuss later on for the user.

It is also worth noting that some older applications used to get away with hiding settings (or even data in the form of binary blobs, and so on) in the registry. This practice was never really a good one to follow, and now it comes with a penalty. Areas of the registry managed by the OS tend to not keep custom key entries when the OS install is upgraded. There is no guarantee that oddball registry hacks from legacy or internal applications are going to migrate for you if they are in registry areas reserved for the operating system as a general rule. Your mileage may vary.

Security mitigation

For the significance of Windows 11's security focus, one simply needs to look at the news. It seems every day that another story emerges of a company or organization that has had ransomware installed and then been blackmailed into paying for an encryption key to regain access to their own data. A review of the work needed to protect from these types of attacks is worth the time.

Additionally, software products working in tandem with antivirus solutions, such as **data loss prevention (DLP)** software or even **intrusion detection software/systems (IDS)** can be used to protect organizations and their data from accidental or even intentional theft by third parties or rogue employees. The typical goal of an organization is to prevent their data from ending up on *WikiLeaks*, so any steps that can be taken toward that end are a good target for the enterprise administrator.

While prevention is all well and good, what about the aftermath of a detected intrusion? Are you prepared for that scenario? More so, is your security team prepared? Forensics tools, Windows log configuration, and subsequent auditing can go a long way toward answering the questions of what happened, how it happened, and what we lost.

With Windows 11, suffice it to say that Microsoft has made many improvements in preventing attacks from occurring. These are discussed in depth in *Chapter 8, Windows 11 Security*.

Internet Explorer 11 retired

Microsoft Internet Explorer 11 reached **end of life (EOL)** in June 2022. Edge Chromium is now the browser of choice with Windows installations (`https://packt.link/F0QPo`).

Windows Store for Business EOL announced

The Windows Store for Business is being retired. There are some plans to make the native Windows 11 store more manageable for enterprise customers, but those details are not yet clear at the time of writing. WinGet is a command-line and scriptable method of doing this; a webinar on this is located here: `https://packt.link/cJLzm`.

> **Note**
> Windows Store for Business will be retiring in the first quarter of 2023.

There is an announcement (and hopefully, more details will become available later on) at this location: `https://packt.link/oeXLI`.

Windows 11 Start and taskbar layout

In Windows 10, there was a lot of difficulty with the **Start** menu configurations. Eventually, there was a PowerShell cmdlet to export and import **Start** menu layouts. Typically, this is done as part of a deployment task sequence using SCCM or MDT to ease the automation of the process. Group Policy and MDM policies can be used to do some of this as well. A good walkthrough of this process is located here: `https://packt.link/CxIp2`.

Do note that this is not supported if you use roaming user profiles.

Audit mode

Audit mode is another method of customizing the default user profile (administrator) for a system. It is a tried-and-true method of manual customization when automation will not fit the situation. One important item of note is that while it is still supported and fine to use, audit mode is not intended or supported as a method of customizing or tweaking the build from upgrade to upgrade. Again, fall back to Group Policies/Group Policy Preferences, and you will be fine here.

Tips

Microsoft has been paying attention to how people use Windows. The constant iteration of the **Start** menu changes from Windows 7 to 8, to 10, and now 11 shows an attempt to streamline application launching and discovery, to reduce the old mechanical process of clicking a **Start** menu tree 5-7 times just to launch an application.

Windows key + *X* is one of the best examples of the work Microsoft has done to optimize the UX and make it more efficient:

Figure 3.2 – Windows key + X dialog

Look at the options available. Most administrative tools can be opened with a simple key combination and a click. This is great!

One thing we have not discussed so far is the usage of Microsoft Intune to help organize and patch devices. Some of the things people want to modify are now managed via MDM. Microsoft Intune is a great way to push MDM settings to enterprise devices and also integrates with on-site SCCM environments.

If you choose to use this tooling, the settings that can be manipulated and managed are documented at `https://packt.link/KO3V6`.

Virtual Desktop Infrastructure

In virtual desktop configurations (where many guest Windows installations reside on a virtualization stack and users connect to them via thin clients or RDP apps), administrators are likely familiar with the variety of scripts used to tweak Windows 7 to make it a performant guest in a **Virtual Desktop Infrastructure** (**VDI**). The scripts were designed to reduce unnecessary I/O load on the disk subsystem of the VDI host(s) as well as reduce CPU usage (except when needed, of course). These scripts made significant changes to the operating system and were supported to varying degrees by vendors, OEMs, and Microsoft.

Provided the admin uses the vendor tool for their VDI (such as the Citrix or VMware VDI optimization tools), performance and stability are both achievable goals. Microsoft also provides a tool that is worth investigating at `https://packt.link/1AqvZ`.

Layering technologies

If one is set on the VDI route, I would suggest exploring layering technology as a fit to bridge the gap of need/capability in Windows. Citrix App Layering is a great example of this capability. These technologies treat the OS image as a layer upon which registry changes, applications, documents, and so on can all be layered into the image before it is presented to the end user.

This thins the data that is relevant to the user considerably when we consider things such as backups, data integrity, and so forth. It also allows enterprises the agility to modify or remove/add applications quickly to a user or group of users, with little of the traditional imaging overhead common to VDI.

SCT

For those concerned with security, Microsoft has had the SCT for some time. This tool lets you take trusted secure baseline configurations from `https://packt.link/OxKKv`, Microsoft, and others and make them into Group Policies that you can import into your environment. Generally speaking, using this tool to securely configure your environment is preferred rather than going off into the woods on your own. The reasons for this are set out here:

- The guidelines are created by expert security entities and professionals.

- When you have trouble and have to get support, is it better to say *We followed the SCT template for secure desktops* or *We did a bunch of tweaks to the registry and security settings and now it doesn't work*? The list of baselines is pretty comprehensive (Windows 11 is in the works at the time of writing and is available at `https://packt.link/Mb2GF`).

> **Note**
>
> For more information about SCM and its implications on security profiles, consult *Chapter 8, Windows 11 Security*.

AppLocker and Windows Defender Application Guard (previously MDAC)

AppLocker is an extension of the native Group Policy software restriction policies. It can be used to block applications wholesale or can be granular, where it will only allow applications to run when they are a particular version or signed with an accepted digital signature/certificate.

Setting up AppLocker is a simple exercise in the Group Policy management console. You can even put all your allowed programs into a `reference` folder and let AppLocker inventory the folder and develop a policy based on those binaries. This is an exercise well worth the effort for administrators looking to prevent malware in their environment.

Windows Defender Application Guard (**WDAC**) is a newer method of managing application presence and usability. Applications security is managed via policy, either GPO, Intune, or other script-based solutions. Depending on the version of Windows, either WDAC or AppLocker is the right solution. The application control capabilities of both solutions are discussed here: `https://packt.link/tz409`.

> **Note**
>
> For more information on AppLocker, refer to *Chapter 8, Windows 11 Security*.

Microsoft telemetry

The advent of forced telemetry in Windows 10 caused a stir in the IT pro and enterprise administration space. For those unaware of this, Windows 10 and 11 keep logs of many activities performed on them and ship those (anonymized) data points back to Microsoft for advanced analytics. Before you panic, let us explore what is collected and why.

What is collected? Let's take a look:

- The type of hardware being used

- Applications installed and usage details

- Reliability information on device drivers

Why is it collected?

Microsoft gives many reasons for collecting this data. The general takeaway here should be that Microsoft uses telemetry to do its best on the functionality of future versions, as well as spending the resources to fix problems in a real-world priority scenario. For example, in the past, if 10,000,000 crashes occurred in `Explorer.exe` daily in the world and they all had the same debugging call stack in them, Microsoft might not have really been aware of this issue until either many calls were made by end users at home or enterprise customers called in with some frequency on the issue.

With Windows 11, Microsoft is listening to the stability metrics of the code it writes. Given the same 10,000,000-crashes-a-day scenario in Windows 11, you can rest assured that Microsoft would dedicate resources to address the problem with all due haste. So, there is a benefit here for home users, enterprise users, and everyone in between.

Now given all this, can you opt out? If you are a home user, not really, no. If you are an enterprise or school user and are using the appropriate license/SKU, then yes, you can. But should you? Does the *potential* loss of *important* data to Microsoft or third parties outweigh the benefits to all users (including your organization) having a better experience? For some organizations, this is an easy decision tree. For others, certainly, it may be a more complicated scenario.

There are different levels of telemetry collected as well. They go from a baseline of security collections only up to a full-blown delivery of application usage patterns at the highest level. Given the changing nature of the WaaS model, I encourage you to review the whole concept of telemetry as it exists during your implementation process. Currently, the telemetry settings are documented in depth at `https://packt.link/NkJ8f` and are worth looking at to understand the exposure (if any) versus the gain.

Windows Spotlight

Windows Spotlight was a new feature in Windows 10 that allows you to have more than just an image for your lock screen. Instead of just a static page, you can now tweak (as a user or as an enterprise administrator) two items, as follows:

- Which image(s) can appear as lock screens?

- Does Windows also display random tips and tricks to you on your lock screen?

Most organizations configure the lock screen to be a corporate logo or corporate-approved art pack to avoid HR issues from occurring and to create uniformity in the office.

The tips most people can take or leave. I find most enterprises turn them off just in case a tip directs the user to do something the company does not want them doing (such as trying to self-resolve an issue rather than contacting the help desk for assistance).

Group Policy can manage the settings for this capability in the enterprise, and that is the recommended method of managing it.

Mandatory user profiles

Mandatory user profiles have been around for some time now, since Windows XP, in fact. For those not familiar with this venerable Windows mechanism, mandatory user profiles are roaming profiles that have been configured with specific settings that are typically not able to be modified by the end user logging on to the Windows machine. Further, any changes to the profile that do get made (for example, malware) are not saved back to the mandatory profile. They are a one-way street of configuration. These are great for education machines: testing centers, writing labs, and kiosks sometimes fit a mandatory profile requirement.

When a server hosting the mandatory profile is unavailable (network issues, remote host away from the corporate LAN, and so on), a locally cached copy is loaded (if it exists, this is configurable). If the profile is not cached locally, a temporary profile can be served or the login can be rejected (via Group Policy).

Mandatory user profiles are, by and large, normal user profiles; `Ntuser.dat` has just been renamed to `.man` (for mandatory), marking the profile read-only. The process is documented in detail on `https://packt.link/7G27C`, so we won't repeat it here (and it is, in theory, subject to change anyway from build to build in the WaaS model).

One concern of mandatory profiles is login times. If you thought copying a profile across the network from a central host (even DFS or other replication) would make the user wait, you are in fact correct. It does. Also, a poorly configured mandatory profile (or even roaming profiles that aren't mandatory) can be a huge cause of **Slow Boot, Slow Logon** (**SBSL**) problems in the enterprise. Microsoft has provided this policy grid to demonstrate which policy functions to use depending on the version of Windows:

Apply policies to improve sign-in time

When a user is configured with a mandatory profile, Windows 10 starts as though it was the first sign-in each time the user signs in. To improve sign-in performance for users with mandatory user profiles, apply the Group Policy settings shown in the following table. (The table shows which operating system versions each policy setting can apply to.)

Group Policy setting	Windows 10	Windows Server 2016	Windows 8.1	Windows Server 2012
Computer Configuration > Administrative Templates > System > Logon > **Show first sign-in animation** = Disabled	✓	✓	✓	✓
Computer Configuration > Administrative Templates > Windows Components > Search > **Allow Cortana** = Disabled	✓	✓	✗	✗
Computer Configuration > Administrative Templates > Windows Components > Cloud Content > **Turn off Microsoft consumer experience** = Enabled	✓	✗	✗	✗

Figure 3.3 – Group Policies to improve sign-in time

Hopefully, some of these policy options are viable for your build to help with sign-in times.

Assigned Access, also known as kiosk mode

I've mentioned kiosk functionality a few times; as it turns out, Windows 11 comes with a feature that will turn your enterprise build into a kiosk serving a single application. So, to do this manually, go to **Settings | Accounts | Other people | Set up assigned access**. Note that multi-app kiosk mode is now slated for 23H2, to be released in the next few weeks at the time of writing.

From here, it is as simple as assigning an account and an application that the account runs (essentially as its shell):

Figure 3.4 – Assigned Access dialog box

Once this is assigned an account and an application, when the user logs in, it opens that application. If the application closes, the user logs out.

For enterprise management, however, doing this configuration individually will just not scale. So, there are guides on `https://packt.link/7G27C` on how to use PowerShell to configure this as well as MDM policies or even the Windows **Imaging and Configuration Designer** (**ICD**).

Windows libraries

Windows libraries have come a long way since their inception. We're at the point now where they can easily include features such as federated search, indexing, and searching for media that are on servers or home computers. There is a lot of flexibility here for the enterprise to present corporate data assets in logical methods other than "*My data is on G:*" and so forth.

You can even implement folder redirection for known folders in libraries. It's important to tread carefully here, though, as slow performance can be encountered with folder redirection implemented badly. The capability of a central rollout of library configuration is done with a library description file and is managed in an XML schema file.

There are still some restrictions in place: no files hosted in Microsoft Exchange or Microsoft SharePoint, no files on NAS devices, and no DFS-hosted files.

Summary

As you can see, Windows 11 brings a lot to bear for enterprise administrators. But it is, again, a paradigm shift from the old Windows 7 image-crafting days. Carefully evaluate the capabilities at your disposal prior to starting your migration and adoption of this new technology, if possible.

In the next chapter, the administration of user accounts will be discussed, including local, domain, and Azure domain-joined accounts.

4

User Account Administration

In this chapter, we will cover the concepts and technologies that enable the secure and productive use of the Windows 11 operating system, as well as the advanced options available to secure the user account credentials and prevent unauthorized system configuration changes and software installation.

We will explore the following topics:

- Windows account types
- Account privileges
- Local Administrator Password Solution
- Creating policies to control local accounts
- Managing user sign-in options
- User Account Control
- Privileged Access Workstation

Windows account types

The Windows 11 operating system supports five types of accounts, each used to enable different functionalities:

- **Service account**: These accounts are used to run background services and are assigned specific permissions. They are not used to log in to the system but may be used remotely. Domain-joined computers may have additional service accounts assigned to enable central administration.

- **Local user account**: By default, at least one local user account is created to run as the local administrator when first configuring the operating system. Depending on how Windows is installed, this account may be a generic account, such as an administrator, or it could be named after the first user who completes the first-time run wizard and chooses not to register a Microsoft account. These accounts are governed by the local password policies, which can be configured via Group Policy or a device/application management service such as **Microsoft Intune**.

- **Microsoft account**: If the computer is not domain-joined, the user can register their Microsoft account (such as @outlook.com) as their local user account. In this configuration, all user settings are synchronized with the Microsoft cloud to provide a seamless transition between multiple computers, or when rebuilding the computer. Microsoft accounts can coexist with local user accounts and Azure **Active Directory** (**AD**) accounts.

- **Azure AD user account**: This account type has the user's corporate credentials stored in Azure AD, such as an **Office 365** user. This login method can be enabled in one of two scenarios:

  - If the computer account is joined to Azure AD (also known as **Workplace Join**), then the user can sign in with their corporate credentials in Azure AD.

  - If the computer account is not joined to Azure AD, the user can sign in with either a local user account or a Microsoft account and then link their Azure AD account using the **Connect to work or school** option. When this is done, the user will be able to store their credentials securely to enable **single sign-on** (**SSO**) to company applications such as Office 365.

- **Windows Server AD user account**: Windows 11 Enterprise computers not joined to Azure AD are likely either to be joined to a Windows server Active Directory domain or Azure Active Directory. When this occurs, the Microsoft account and Azure AD user account options are disabled. However, the AD user account can be automatically linked to the Azure AD user account, to enable SSO when the user is not on the corporate network.

Account privileges

Each account can be assigned a range of specific privileges, from a standard user account (with no systems access) to a full local administrator account. Gaining access to administrative rights on the Windows operating system is one of the key attack vectors that needs to be prevented in every organization and even on personal PCs. Administrative rights are required when changing configurations or installing software, both of which should not be carried out by users, and therefore all user accounts should be restricted to standard user accounts only.

Where there is a genuine need for a user to be granted local admin rights on a computer, they should never be assigned to the user's main account that they use for gaining access to email, documents, and websites. This leads to the potential for a user to open a document, or click on a hyperlink, that contains malware. A better design approach is to create a local user account specific to this user and provide them with the password to the account. This way, when the user needs to make a change or install software, they can enter the local admin user ID and password into the **User Account Control** (**UAC**) prompt. Alternatively, endpoint privilege management (part of Intune) may be configured.

For Azure Active Directory or AD domain-joined computers, the local administrator accounts should not share the same user ID and password. Instead, **Local Administrator Password Solution** (**LAPS**) should be used. Domain-joined computers will have specific domain accounts added to the local permissions on every PC, such as the local administrators group. This allows IT administrators and management systems to connect remotely to help support and configure the computers. IT administrators should not log on to any computers with domain-level privileged accounts (such as domain admin); this is one of the most well-known methods of gaining domain-wide permissions through pass-the-hash attacks.

Instead, they should carry out their work via **PowerShell** remotely, and if they need to log in interactively, they should take over the user's session and use the **Run as administrator** option (obtaining the current unique password for the local administrator account first using LAPS).

Local Administrator Password Solution

If a single password is configured for the local admin accounts across all domain-joined computers, there is a high risk that it can be used in a widespread attack to install malware, elevate privileges, or gain access to sensitive files. To resolve this issue, Microsoft offers the **Local Administrator Password Solution** (**LAPS**). This works by setting a different random password on every computer in the domain and storing that password in AD, or Azure AD if it is used. Administrators can choose who can access those passwords in order to support the PCs.

The solution is built into AD and doesn't require any other supporting technologies or licenses. LAPS uses the Group Policy **client-side extension** (**CSE**) or CSP that you install on managed computers to perform all management tasks. The solution's management tools provide easy configuration and administration.

Once configured, you can create Group Policy settings to enable local administrator password management and control the configuration of the password settings, as follows:

- **Password settings**: Some default settings are already configured (see `https://packt.link/KFQDj` for reference):

  - Characters used when generating a new password:

    - Default: Capital letters + small letters + numbers + special characters

  - Password length:

    - Minimum: 8 characters

    - Maximum: 64 characters

    - Default: 14 characters

  - Password age in days:

    - Minimum: 1 day

    - Maximum: 365 days

    - Default: 30 days

- **Name of administrator account to manage (optional)**:

  - The name of the local account you want to manage a password for:

    - Do not configure when you use a built-in admin account. A built-in admin account is auto-detected by a well-known SID, even when renamed.

    - Do configure when you use a custom local admin account.

- **Do not allow password expiration time longer than required by policy**:

  - When you enable this setting, planned password expiration longer than the password age dictated by the password settings policy is not allowed. When such expiration is detected, the password is changed immediately and password expiration is set according to policy.

  - When you disable or don't configure this setting, the password expiration time may be longer than required by the password settings policy.

- **Enable local admin password management**:

  - This enables password management for a local administrator account

  - If you enable this setting, the local administrator password is managed

  - If you disable or don't configure this setting, the local administrator password is *not* managed

At the next Group Policy refresh, the passwords are changed. Authorized administrators can then use the LAPS UI tool to search individual computers to retrieve the password and/or change the expiry date for the next password change (see *Figure 4.1*):

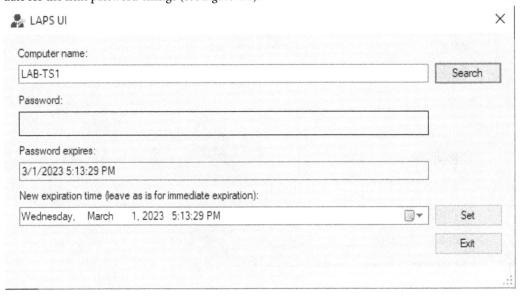

Figure 4.1 – LAPS administration panel

You can download the tools and configuration details here: `https://packt.link/xg3tB`

Creating policies to control local accounts

If you enable local admin accounts, for users who require them, you should also enforce a set of policies to ensure the local accounts have strong authentication standards. On domain-joined computers, Group Policy can be used to specify the settings of the local account policy, which contains two subsets:

- **Password policy**: These policy settings determine the controls for local account passwords, such as enforcement and lifetimes
- **Account lockout policy**: These policy settings determine the circumstances and length of time for which an account will be locked out of the system when the password is entered incorrectly

Password policy

The password policy enforces specific values that control how often the password is changed, how complex it is, and whether users can reuse old passwords. The default values are shown in the following screenshot:

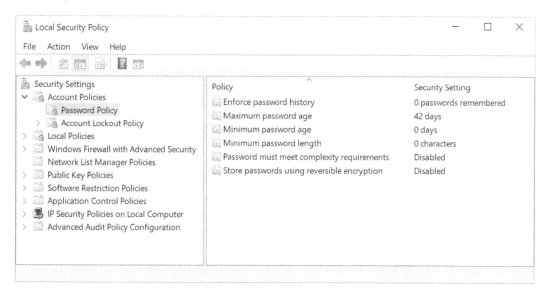

Figure 4.2 – Local Security Policy's password policy

You may want to configure this policy to be more restrictive than the domain-level password policy. The following table provides some recommendations for each of the values:

Policy	Recommended setting	Justification
Enforce password history	24	This makes it less likely that a user will attempt to reuse the same password
Maximum password age	30 days	Passwords for privileged accounts should be changed on a frequent basis
Minimum password age	1 day	This setting ensures the password is not reset multiple times in one day to return to a favorite password
Minimum password length	14	As this account is not expected to be used very often, the usability of the password should be less of a concern compared to security against password-cracking attempts
Password must meet complexity requirements	Enabled	Even a 12-character password should contain complexity, such as uppercase, lowercase, numeric, and special characters
Store passwords using reversible encryption	Disabled	This setting should never be enabled

Table 4.1 – Password policies

Azure Active Directory maintains its own defaults, which are shown in the following table:

Property	Requirements
Characters allowed	A – Z a - z 0 – 9 @ # $ % ^ & * - _ ! + = [ ] { } | \ : ' , . ? / ` ~ " ( ) ; < > Blank space
Characters not allowed	Unicode characters

Property	Requirements
Password restrictions	A minimum of 8 characters and a maximum of 256 characters. Requires three out of four of the following types of characters: • Lowercase characters • Uppercase characters • Numbers (0-9) • Symbols (see the previous password restrictions)
Password expiry duration (maximum password age)	Default value: 90 days. If the tenant was created after 2021, it has no default expiration value. You can check the current policy with `Get-MsolPasswordPolicy`. The value is configurable by using the `Set-MsolPasswordPolicy` cmdlet from the Azure Active Directory module for Windows PowerShell.
Password expiry (let passwords never expire)	Default value: false (indicates that passwords have an expiration date). The value can be configured for individual user accounts by using the `Set-MsolUser` cmdlet.
Password change history	The last password can't be used again when the user changes a password.
Password reset history	The last password can be used again when the user resets a forgotten password.

Table 4.2 – Azure Active Directory property and requirement

Account lockout policy

If an attacker attempts to guess the password of a local administrative account, the lockout policy will slow down their attempts by enforcing further restrictions on the number of attempts that can be made in a set time period. This, combined with the increased complexity of the password, should make it very difficult for a successful attack to take place before the account password expires.

The default values are shown in the following screenshot:

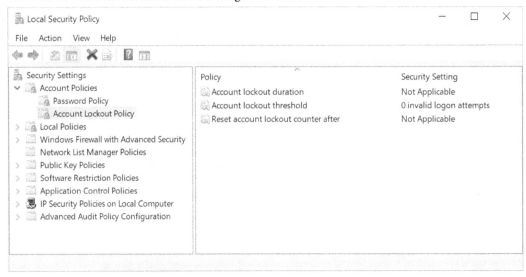

Figure 4.3 – Local Security Policy's account lockout policy

You should configure this policy to be more restrictive than the defaults that are set. The following table provides some recommendations for each of the values:

Policy	Recommended Settings	Justification
Account lockout duration	10 minutes	Once the maximum number of password attempts is reached, the account is locked for 1 hour before further attempts can be made
Account lockout threshold	10 attempts	A genuine user may make mistakes when entering a complex password, but they should be expected to enter the correct password within 15 attempts
Reset account lockout counter after	10 minutes	This specifies the time period that the account lockout threshold monitors (24 hours)

Table 4.3 – Account lockout policies

In summary, the preceding configuration will allow an attacker to make no more than 15 bad password attempts every 24 hours; if they do, the account is locked for 1 hour every time a subsequent bad attempt is made. This increases the usability for a genuine user who may make several bad attempts (but not 15) without being locked out unnecessarily, while also making it extremely difficult for a brute-force attack to be achieved before the password is changed again.

Managing user sign-in options

Windows 11 Enterprise offers a range of configurable options to manage the account logon process. Some of the features are designed to increase security, while others are to improve the user experience. It's worth mentioning that with **Multifactor Authentication** in use, the password strength for a user is less of a bulwark for enterprise security.

The following settings can be configured via GPO to ensure a consistent approach across all domain-joined computers:

- **Turn on convenience PIN**: This setting should be disabled as it causes the password to be cached in the system vault; instead, use the **Hello for Business** feature that we will see later in this chapter.

- **Turn off picture password sign-in**: This policy should be enabled to prevent the use of this feature. Picture password sign-in enables the user to sign in with a unique gesture based on their picture, but also causes the user's password to be cached in the system vault. Windows Hello for Business is a better option.

- **Do not enumerate connected users on domain-joined computers**: This setting is generally used to obfuscate the login credentials for domain users who have logged on to the computer. While this may seem a good security measure to prevent an attacker from identifying the user account details, it makes the user experience more difficult as they have to enter their user ID every time they reboot or log off, especially if they are the primary, or only, user of the computer.

- **Enumerate local users**: The default behavior is to not enumerate local user accounts on domain-joined computers. This is the recommended configuration.

- **Block user from showing account details on sign-in**: This policy prevents the user from showing account details, such as email address or username, on the sign-in screen. The default behavior is to allow this, but we recommend disabling this feature.

- **Turn off app notifications on the lock screen**: This setting allows you to prevent app notifications from appearing on the lock screen. While this may be a convenience to the user, displaying notifications on the lock screen can make sensitive data visible to anyone who sees the screen, without the need to log on.

- **Configure Windows Spotlight on lock screen**: If you enable this policy setting, Windows Spotlight will be set as the lock screen provider and users will not be able to modify their lock screens. Windows Spotlight will display daily images from Microsoft on the lock screen. Additionally, if you check the **Include content from Enterprise Spotlight** checkbox and your organization has set up an Enterprise Spotlight content service in Azure, the lock screen will display internal messages and communications configured in that service, when available. If your organization does not have an Enterprise Spotlight content service, the checkbox will have no effect. *Note: This policy is only available for Enterprise SKUs.*

- **Turn off toast notifications on the lock screen**: Again, for privacy reasons, it is recommended this option be enabled to suppress notifications without first signing in.

- **Allow Cortana above lock screen**: This policy setting determines whether or not the user can interact with Cortana using speech while the system is locked. If you enable or don't configure this setting, the user can interact with Cortana using speech while the system is locked. If you disable this setting, the system will need to be unlocked for the user to interact with Cortana using speech. Consider the potential privacy risks associated with this capability.

As you consider the appropriate configuration of the settings for your environment, ensure you find the right balance between usability, privacy, and security. Options that prevent the user from having to enter their password at each login can lead to increased security by removing the temptation for the user to write down their password, or make it so simple that it's easy to guess. Encourage users to adopt these new login methods, along with creating more complex passwords of 12 or more characters as they will not have to enter it every day.

IT administrators need to take special care of where they log in, as compromised credentials can lead to devastating attacks by malicious users. For great guidance on how to mitigate this risk, see the section on **Privileged Access Workstation (PAW)** later in this chapter.

For more Group Policy settings, download the spreadsheet provided by Microsoft: `https://packt.link/zeiTL`

Mobile device management security settings

If you are managing your computers with a **mobile device management (MDM)** solution such as Microsoft Intune, you have the following security settings available:

- **Required password type**: Specifies the type of password that's required, such as alphanumeric or numeric only.

- **Required password type - minimum number of character sets**: Specifies how many different character sets must be included in the password. There are four character sets: lowercase letters, uppercase letters, numbers, and symbols. However, for iOS devices, this setting specifies the number of symbols that must be included in the password.

- **Minimum password length**: Configures the minimum required length (in characters) for the password.

- **Number of repeated sign-in failures to allow before the device is wiped**: Wipes the device if the sign-in attempts fail this number of times.

- **Minutes of inactivity before screen turns off**: Specifies the number of minutes a device must be idle before a password is required to unlock it.

- **Password expiration (days)**: Specifies the number of days before the device password must be changed.

- **Remember password history**: Specifies whether the user can configure previously used passwords.

- **Remember password history - prevent reuse of previous passwords**: Specifies the number of previously used passwords that are remembered by the device.

- **Allow picture password and PIN**: Enables the use of a picture password and PIN. A picture password lets the user sign in with gestures on a picture. A PIN lets users quickly sign in with a four-digit code. While these are useful options for the user, the better approach is to disable these features and use Windows Hello for Business instead (we'll discuss it later in this chapter).

Reference
https://packt.link/B77SM

User Account Control

User Account Control (UAC) is a fundamental security control that helps mitigate the impact of malware, yet some enterprise administrators disable UAC at the request of the users because it is seen as an annoying and unnecessary prompt that gets in the way of productivity. The feature has improved greatly since it was first launched (as part of Windows Vista), so we encourage you to ensure this is enabled across all managed computers in your environment. Microsoft tests all software with the defaults enabled, therefore disabling UAC may cause unexpected results for application launches or security configurations.

With UAC enabled, Windows 11 prompts for consent, or for credentials of a valid local administrator account, before starting a program or task that requires a full administrator access token. This prompt ensures that no malicious software can be silently installed.

If the user is logged on with local admin rights (which is not recommended), the consent prompt is presented when a user attempts to perform a task that requires a user's administrative access token. The following is an example of the UAC consent prompt you will see if you have local admin rights.

Figure 4.4 – UAC prompt

Alternatively, the credential prompt is presented when a standard user attempts to perform a task that requires administrative access, such as installing software or making a system configuration change (both potential signs of malware). Administrators can also be required to provide their credentials by setting the **User Account Control: Behavior of the elevation prompt for administrators in Admin Approval Mode** policy setting value to **Prompt for credentials**.

The following is an example of the UAC credential prompt:

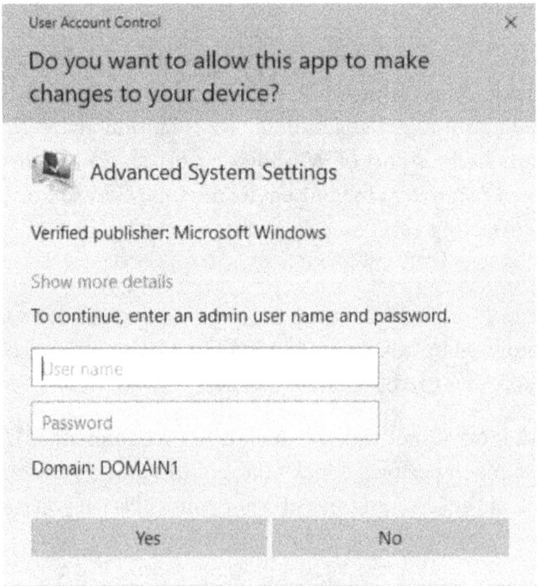

Figure 4.5 – UAC authentication prompt

Other settings that can be controlled by UAC are as follows:

- **Admin Approval Mode for the Built-in Administrator account**: Controls the behavior for the built-in administrator account only. We recommend this setting to be enabled.

- **Behavior of the elevation prompt for administrators in Admin Approval Mode**: Options include prompting for consent, prompting for credentials, or elevating without prompting. We recommend this setting be configured to prompt for consent or credentials.

- **Behavior of the elevation prompt for standard users**: Options include prompting for credentials or automatically denying elevation requests. If the user is not provided with a separate administrator account, then set this value to automatically deny (the default behavior for Enterprise).

- **Detect application installations and prompt for elevation**: This setting determines the behavior of the entire system. Options include prompting for elevation (consent or credentials) and disabling. The default behavior for Enterprise is set to **Disabled** because managed software does not require the user to have local admin rights to install.

- **Only elevate executables that are signed and validated**: If enabled, this security setting enforces **public key infrastructure** (**PKI**) signature checks on any interactive application that requests elevation of privilege.

- **Only elevate UIAccess applications that are installed in secure locations**: This option can be used to enforce the requirement that applications that request execution with a **User Interface Accessibility** (**UIAccess**) integrity level must reside in a secure location on the filesystem.

- **Run all administrators in Admin Approval Mode**: This security setting determines the behavior of all UAC policies for the entire system. We recommend this be set to **Enabled**.

- **Switch to the secure desktop when prompting for elevation**: Secure desktop provides a clear indication to the user that elevation is being requested, or the prompt may be hidden behind other windows. Disabling this option increases your security risk, so it is recommended this is set to **Enabled**.

- **Virtualize file and registry write failures to per-user locations**: Virtualization facilitates the running of pre-Vista (legacy) applications that historically failed to run in the standard user context. As you are deploying Windows 11 Enterprise, it is very unlikely you will have applications that still require this configuration.

- **Allow UIAccess applications to prompt for elevation without using the secure desktop**: UIAccess programs are designed to interact with Windows and application programs on behalf of a user. This setting allows UIAccess programs to bypass the secure desktop to increase usability in certain cases, but allowing elevation requests to appear on the regular interactive desktop instead of the secure desktop increases your security risk. We recommend disabling this option.

For more information on how UAC works, see here: `https://packt.link/DKrSs`

Privileged Access Workstation

If you really want to take security seriously, then you need to provide the highest levels of security for your privileged accounts to prevent malicious behavior through compromised access. Microsoft has developed a complete set of guidance materials on how to configure specific workstations used by administrators and other privileged accounts to carry out sensitive tasks, such as systems administration and high-value financial transactions.

In this model, the computers are designated specifically for privileged access, blocking any other accounts from logging on interactively or via the network. Instead of logging on to the computer as a standard user and elevating privileges to gain access to sensitive information and systems, the user logs onto the PAW computer directly with the privileged account and carries out the tasks required.

This system works by preventing the usual risky behaviors such as internet browsing, opening emails and attachments, or running unsanctioned programs from occurring on a computer. By locking down the computer to only run verified and trusted applications and only allowing a minimal set of trusted accounts to gain access, the computer remains as secure as possible. Other systems can then be configured to only allow administrators to log in if the request originates from a PAW computer, and to deny all other administrative login attempts.

The guide provides a phased approach to deployment, ensuring you can quickly gain the benefits and protect the most critical accounts: domain administrators. Once that level of protection is in place, you can extend it to other privileged accounts and configure security for administering cloud services such as Office 365 and Azure.

You can find out more here: `https://packt.link/4vNPi`

Summary

Windows 11 Enterprise provides the tools required to deliver a secure environment to access sensitive and valuable information and systems.

There are many options to consider when creating and securing local user accounts that will gain authorized access to your systems. The most important rule is to never log in to computers with local admin rights enabled, and to instead use run-as to elevate rights with a separate administrative account. Also, never log in to a client computer with domain-privileged accounts, and limit logging on to trusted IT PCs only, such as PAWs. Finally, ensure all administrative account passwords are unique across computers, complex, and changed regularly.

In the next chapter, we will explore remote administration for troubleshooting and remote assistance.

5
Tools to Manage Windows 11

IT professionals need ways to help them manage Windows 11, whether in a small shop or an enterprise. The Microsoft **Remote Server Administration Tools** (**RSAT**) is an essential set of tools for managing remote or non-UI servers, a common scenario environment with data centers and increasingly cloud resources. You can also use other community-created tools, including those produced by Sysinternals.

RSAT provided by Microsoft includes GUI applications, PowerShell commands, and Windows console applications. You can install any or all of the tools. PowerShell is increasingly the preferred and advised remote administration method for servers and clients.

The community has also produced many tools to help you manage Windows 11. These specifically include a range of tools from Sysinternals, a leading vendor of Windows-based administration tools. You can support remote servers and workstations with PowerShell and the GUI. These both simplify your administration and enable you to develop automation.

In this chapter, we'll learn about the following topics:

- How to install RSAT
- How to perform administrative tasks using RSAT
- Obtaining and using Sysinternals tools
- Using Sysinternals PsTools for remote management
- Utilizing BGInfo to customize the screen
- Utilizing the PsTools suite of console applications

RSAT

Microsoft has a long history of producing tools to enable you to manage Windows components, particularly those that run on Windows Server (such as DNS, the **Internet Information Services** (**IIS**) web server, Windows clustering, and many more).

With each Windows Server feature, Microsoft has developed GUI and PowerShell modules that help you manage the feature. The primary goal of these tools is to enable you to manage your Windows Server and Active Directory infrastructure from the Windows client, Windows 11. Most of these are mature tools that Microsoft has not enhanced for several years.

Installing the RSAT tools

In early versions of Windows 10, the RSAT tools were a separate download from Microsoft. With the October 2018 Windows 10 update, the RSAT tools became **Features on Demand** (**FODs**) that you can install directly into Windows. This process does require internet access to download the appropriate tools.

There are two ways to install the tools in Windows 11, as we'll see next.

The first way to add the RSAT tools is to use the **Settings** app. Click on the **Start** menu on the Windows 11 taskbar (or use the Windows key + *I*) to bring up the **Settings** menu, then click **Apps** in the left pane, like this:

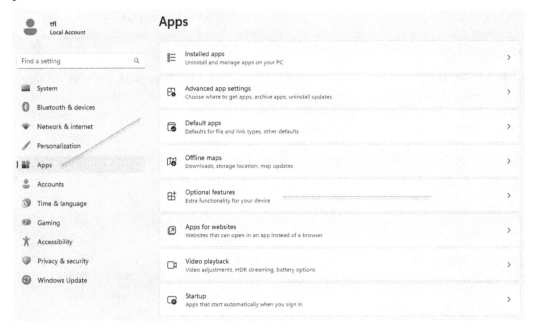

Figure 5.1 – Apps settings pane in Windows 11

Next, click on the **Optional features** button to bring up the **Optional features** pane, which looks like this:

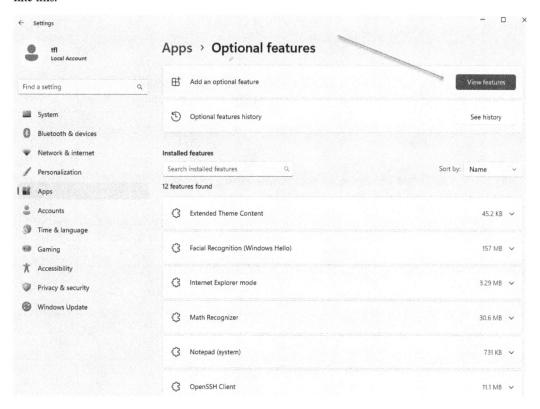

Figure 5.2 – Optional features menu

If you then click on the **View features** button, you can view all the optional features you can now install. To simplify finding the RSAT tools, you can type RSAT in the search field to see all the available RSAT tools, which looks like this:

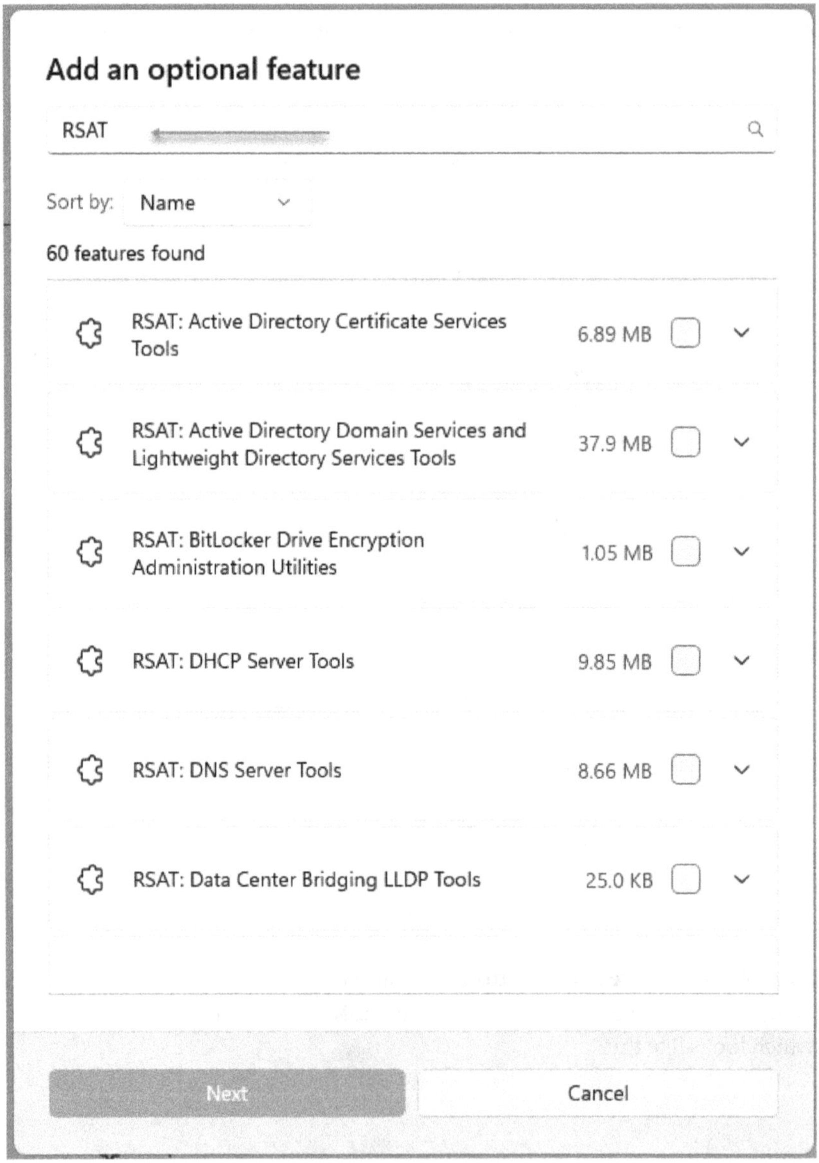

Add an optional feature

RSAT   🔍

Sort by: Name   ⌄

60 features found

🧩	RSAT: Active Directory Certificate Services Tools	6.89 MB	☐ ⌄
🧩	RSAT: Active Directory Domain Services and Lightweight Directory Services Tools	37.9 MB	☐ ⌄
🧩	RSAT: BitLocker Drive Encryption Administration Utilities	1.05 MB	☐ ⌄
🧩	RSAT: DHCP Server Tools	9.85 MB	☐ ⌄
🧩	RSAT: DNS Server Tools	8.66 MB	☐ ⌄
🧩	RSAT: Data Center Bridging LLDP Tools	25.0 KB	☐ ⌄

| Next | Cancel |

Figure 5.3 – Viewing RSAT optional features

To install different RSAT optional extras, click on the tool or tools you wish to add, then click **Next**, like this:

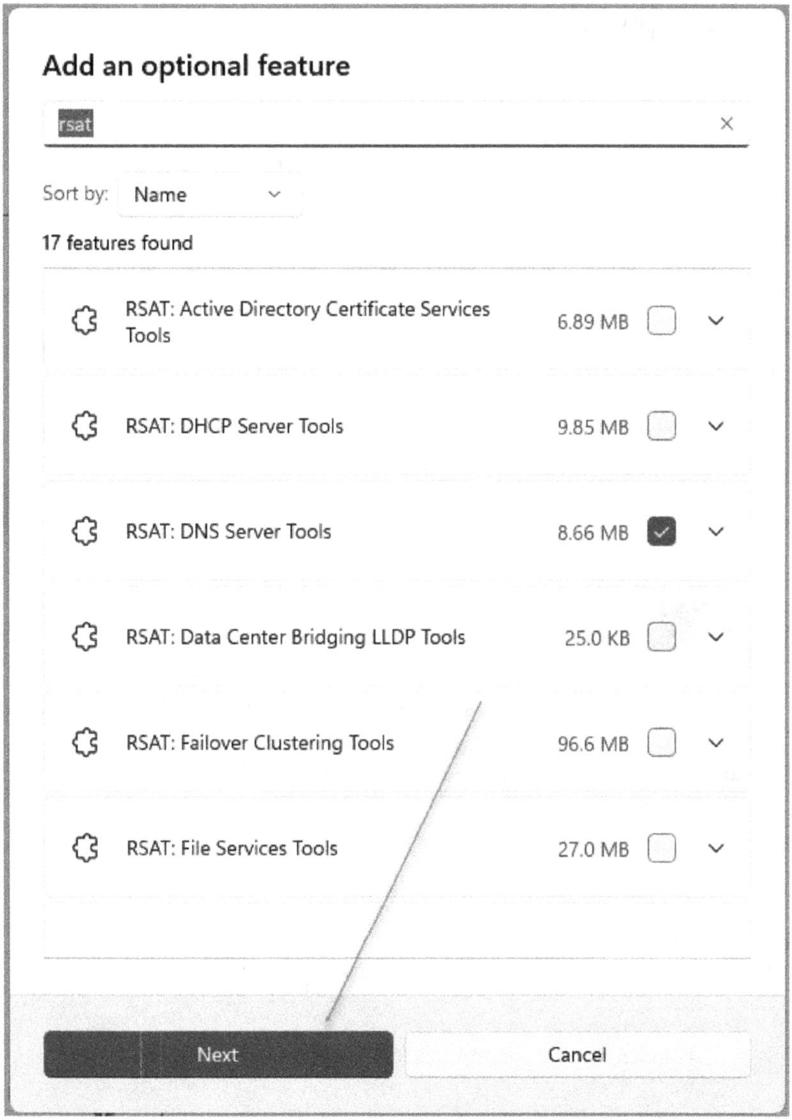

Add an optional feature

rsat   ×

Sort by: Name   ⌄

17 features found

| | RSAT: Active Directory Certificate Services Tools | 6.89 MB | ☐ | ⌄ |

| | RSAT: DHCP Server Tools | 9.85 MB | ☐ | ⌄ |

| | RSAT: DNS Server Tools | 8.66 MB | ☑ | ⌄ |

| | RSAT: Data Center Bridging LLDP Tools | 25.0 KB | ☐ | ⌄ |

| | RSAT: Failover Clustering Tools | 96.6 MB | ☐ | ⌄ |

| | RSAT: File Services Tools | 27.0 MB | ☐ | ⌄ |

| Next | Cancel |

Figure 5.4 – Adding RSAT extras

After clicking **Next**, you'll see a confirmation of the tools you wish to add:

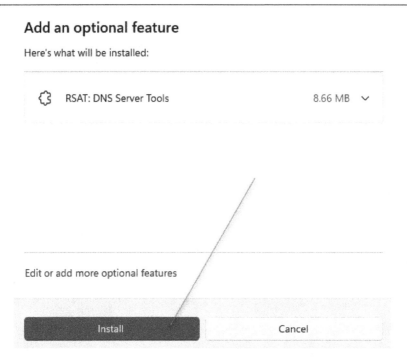

Figure 5.5 – Confirming the addition of DNS Server Tools

After clicking **Install**, Windows downloads and installs the chosen features, such as the DNS Server Tools package. You'll see this:

Figure 5.6 – DNS Server Tools RSAT option being installed

Once Windows completes the installation, you can use either the DNS GUI applet or PowerShell cmdlets.

The second way to install the RSAT tools is to use PowerShell (you read about PowerShell in *Chapter 2, Introduction to PowerShell*). To add RSAT tools to Windows 11, open up an elevated PowerShell console first, and use the Get-WindowsCapability cmdlet to review what is available.

Then, you use the `Add-WindowsCapability` cmdlet to add the chosen tools. If you wanted to add the storage management RSAT option feature, you would do this:

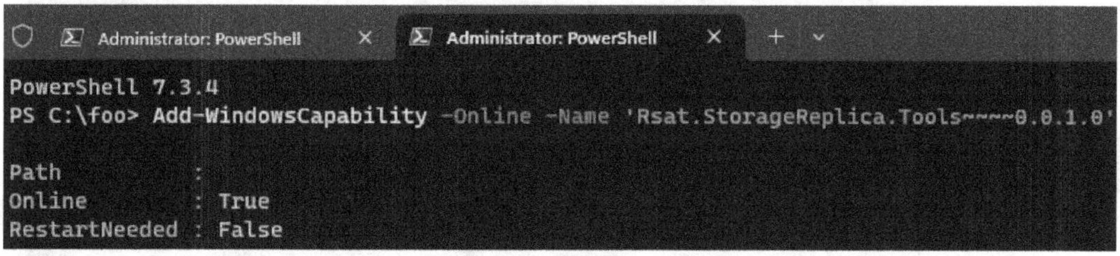

Figure 5.7– Adding an RSAT option using PowerShell

Once you have installed the tools, you need to ensure you keep them up to date.

Updating the RSAT tools

To keep your RSAT tools up to date, you can utilize Windows Update as Microsoft releases updates to address bugs, fix security vulnerabilities, and introduce new features.

The individual RSAT tools contain a variety of commands, GUIs, and so on. PowerShell users should find the RSAT cmdlets are easy to use and follow all the normal PowerShell conventions (noun/verb naming, and so on).

Using the RSAT tools

Each RSAT feature you add typically contains both a GUI tool and one or more PowerShell modules that contain useful cmdlets (and some command-line tools). These tools therefore provide you with the ability to manage Windows Server features/roles using either a GUI tool or PowerShell modules for automation.

The DNS Server RSAT module you installed previously contains 134 different commands you can use with PowerShell (along with the GUI). As noted in *Chapter 2, Introduction to PowerShell*, you can use PowerShell discovery tools (including `Get-Module` and `Get-Command`) to find the RSAT modules and commands they contain. In the case of the DNS Server RSAT module, you can easily discover the commands, as seen here:

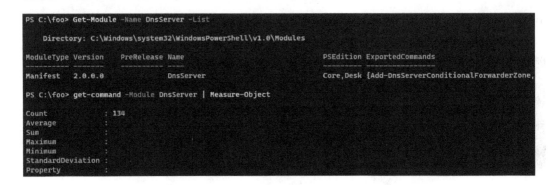

Figure 5.8 – Discovering RSAT commands

Once you have the relevant module, you can use the commands to administer Windows Server components. For example, to discover which zones a DNS server contains, you can use the Get-DNSServerZone cmdlet:

```
PS C:\foo> Get-DnsServerZone -ComputerName cookham1

ZoneName                              ZoneType  IsAutoCreated IsDsIntegrated IsReverseLookupZone IsSigned
--------                              --------  ------------- -------------- ------------------- --------
_msdcs.cookham.net                    Primary   False         True           False               False
0.in-addr.arpa                        Primary   True          False          True                False
1.10.10.in-addr.arpa                  Primary   False         True           True                False
10.10.10.in-addr.arpa                 Primary   False         True           True                False
127.in-addr.arpa                      Primary   True          False          True                False
255.in-addr.arpa                      Primary   True          False          True                False
6.8.3.6.0.1.0.8.2.0.a.2.ip6.arpa      Primary   False         True           True                False
cookham.net                           Primary   False         True           False               False
reskit.org                            Forwarder False         False          False
TrustAnchors                          Primary   False         True           False               False
```

Figure 5.9 – Using the Get-DnsServerZone RSAT option

Now, let's go ahead to our next topic: the Sysinternals tools suite.

The Sysinternals tools suite

Sysinternals is a company founded by Mark Russinovich and Bryce Cogswell. The company developed a powerful set of tools and utilities. You use these tools to manage Windows (Windows Server and Windows 11). In some cases, these tools augment what Microsoft ships and, in other cases, add functionality otherwise missing in Windows. Microsoft acquired Sysinternals in 2006 and has continued to develop these tools, which it makes available for free. You can learn more by navigating to the Microsoft website: https://packt.link/XGaSY.

Downloading the Sysinternals tools suite

You can download the entire suite in several ways or download and use one of the tools.

A simple way to download all the tools is to use your browser and navigate to `https://packt.link/KgkzW`. Then click on the **Sysinternals Suite** link, as shown in the following screenshot:

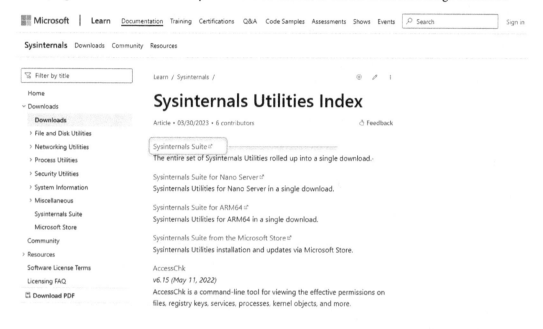

Figure 5.10 – Sysinternals online index

Once your browser has downloaded the ZIP file containing all the tools, you can expand the archive file into a local folder.

If you want to download just a single utility, go to the same download page, and scroll down to find the desired tools—for example, **Autologon**—as shown in the following screenshot:

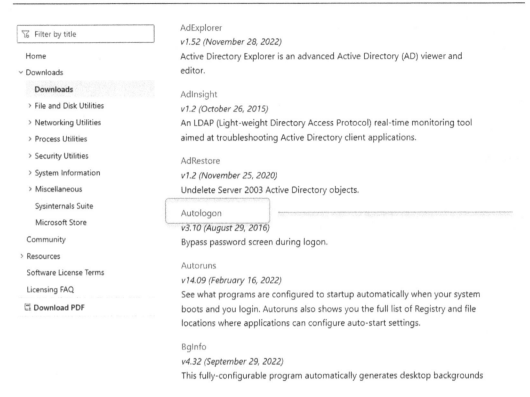

Figure 5.11 – Viewing utilities within the suite

Once your browser has downloaded a single utility, you can run it from your browser's download folder (or move it to another folder).

The Sysinternals suite is also available from the Microsoft Store. Run the Microsoft Store, search for `Sysinternals`, and install the suite. Once installed, you can run it from the installation folder.

> **Important note**
>
> As with most Microsoft Store apps, the MSIX installer installs the Sysinternals suite in a per-user location. However, the installer actually places individual binaries in a secure place, which Windows then shares with other users. You can then run these commands directly from CMD . EXE or PowerShell. For more details on how this works, see https://packt.link/Rx934.

And of course, you can easily download the Sysinternals suite (as a ZIP file) and then expand the download into a more convenient place. The following code downloads the suite from the internet, unblocks the download, and then expands it into a location of your choice:

```
# Set the download URL and target directory
$DownloadUri   = 'https://download.sysinternals.com/files/
```

```
SysinternalsSuite.zip'
$TargetFolder  = 'C:\Sysinternals'
$TargetTemp    = "$TargetFolder\SysinternalsSuite.zip"
New-Item -Path $TargetFolder -ItemType Directory -ea 0| Out-Null
# Download the ZIP file
Invoke-WebRequest -Uri $DownloadUri -OutFile $TargetTemp
Unblock-File -Path $TargetTemp
# Expand the ZIP file
Expand-Archive -Path $TargetTemp -DestinationPath $TargetFolder -Force
# Cleanup to remove the downloaded ZIP file
Remove-Item $TargetTemp
```

Once downloaded, you can run any tool directly from the target folder (that is, C:\Sysinternals). If you do not know the tool suite well, downloading the suite to a known location may be more convenient since the store install process hides the actual applications. Note that this method does not keep the tools up to date, so if Microsoft releases updates or new tools, you have to re-download the tools, as we showed earlier.

Another useful way of obtaining specific utilities is to navigate to https://packt.link/GkoqC. The resulting page displays the full set of tools (and supporting files such as help files), which you can then download. This page enables you to run any Sysinternals tool from any computer connected to the internet without having to navigate to a web page, download, and extract the ZIP file.

The full Syinternals suite consists of a large number of tools. A complete look at the tools would take up too much space in this chapter, so here we concentrate on a few utilities. To learn more about the individual utilities, you can view the index page at https://packt.link/LSLAs. This page has links to more documentation about the utilities. The following sections will explore two specific tools: **BGInfo**, for displaying useful information on a user's desktop, and the **PsTools** suite of remote control tools. The PsTools suite allows you to manage remote hosts.

Introducing the Sysinternals BGInfo tool

BGInfo is a very useful tool available from the Systinternals suite of tools. This tool makes specific system information available when the user logs in by displaying it as text in the background or a pop-up box in the system tray. This information can be very handy when you need to call the helpdesk to report an issue and provide detailed information to help troubleshoot it, especially if you are troubleshooting a VM. You can deploy BGInfo to some or all computers and apply it to the login screen and users' desktop wallpaper—even if the user has a custom background on their desktop that changes frequently, BGInfo overlays the text on top of the custom image.

This section explains how to configure the options you want to display and then deploy the tool and custom configuration to all your computers to ensure a consistent display of information for everyone.

Configuring BGInfo

You start the configuration process by launching the BGInfo tool on your local computer (or one that represents the configuration of your intended targets) as follows:

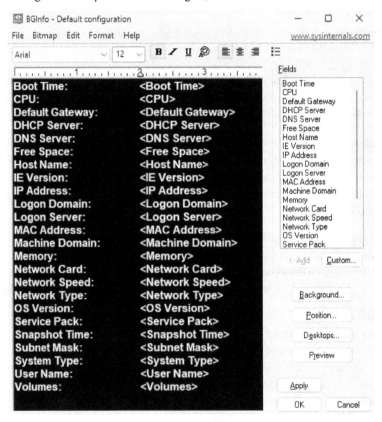

Figure 5.12 – BGInfo tool

You can then configure your BGInfo options to specify what information you wish BGInfo to display. Once you identify the information you wish to display on the screen (and how you want BGInfo to display it), you can click **OK** to set the screen. Alternatively, save the custom BGInfo configuration file (use **File | Save**). By default, the background looks like this:

Figure 5.13 – BGInfo in action

BGInfo allows you to customize further, including which fields to display (and their order), where to display this information, and details about text format. This helps reduce information on the screen, as shown in the following screenshot:

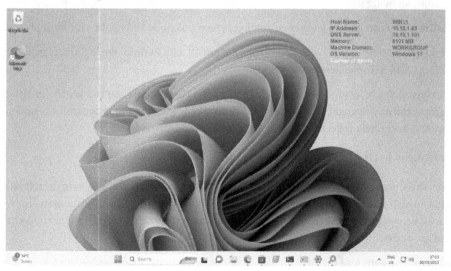

Figure 5.14 – Customizing BGInfo display

From BGInfo, using the **Custom** button allows you to specify which options to customize the display further.

The following options are available to configure what information is to be displayed and how it appears:

- **Fields**: Select what information should be collected and displayed. You can choose from a list of standard options or create your custom fields, as you can see in this screenshot:

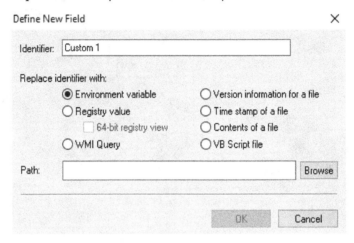

Figure 5.15 – BGInfo custom field definition

- **Background**: Choose a solid color or wallpaper for the background. You can also copy the existing settings, allowing users to select their backgrounds while still displaying BGInfo information.

- **Position**: You can choose from nine possible screen positions and options for multiple screens.

- **Desktops**: Choose which types of desktops are updated by BGInfo: the logon screen, local user desktop, and Remote Desktop logon.

- **Preview**: If you select this button while configuring the options, you can see the image in the background as you create and change the options. This information is dynamically updated to ensure it accurately reflects your configuration options.

The **BGInfo** menu has the following options:

- **File**: This menu allows you to save various copies of the files and restore them to default settings. You can also specify a database or a text file that BGInfo uses to store information. These files can be very useful if you are trying to carry out an audit or check for configuration changes that are occurring.

- **Bitmap**: The output of the BGInfo file is a bitmap image; this menu option allows configuring the size and quality of the image.

- **Edit**: This menu allows you to insert an image such as your company logo.

- **Format**: This menu provides a wide range of options to modify the text's size, style, and color to suit your preference.

You should now have a customized `.bgi` file to configure other computers. The next step will show you how to deploy this to all your computers.

Deploying BGInfo

To deploy BGInfo, you need to determine how to transfer both the BGInfo executable (`.exe`) and the BGInfo configuration file (`.bgi`) to remote computers as well as how you configure running BGInfo at logon. The simplest method is to create a share on a central server that all clients can reach and then upload the two files (`.exe` and `.bgi`). The next step is to create a script and run it on each computer: in an Active Directory environment, create a logon script and deploy it using Group Policy to target the computers.

You can also use system management tools such as **Microsoft Endpoint Configuration Manager (MECM)** to deploy BGInfo.

When you execute `Bginfo.exe`, you can specify several options to modify the behavior of BGInfo, as follows:

- `<path>`: Specifies a configuration file's name for the current session. When you deploy this file, set it to read-only share to ensure other users cannot modify the settings.

- `/timer`: Specifies the timeout value for the countdown timer (in seconds). Specifying zero will update the display without displaying the configuration dialog. Specifying 300 seconds or longer disables the timer altogether.

- `/popup`: Causes BGInfo to create a pop-up window containing the configured information without updating the desktop. The information is formatted just as it would if displayed on the desktop but is contained in a separate window. When using this option, the history database is not updated.

- `/silent`: This option suppresses any error messages; remove it when troubleshooting any issues.

- `/taskbar`: Causes BGInfo to place an icon in the taskbar's status area without updating the desktop. Clicking on the icon causes the configured information to appear in a pop-up window. When using this option, the history database is not updated.

- `/all`: Specifies that BGInfo should change the wallpaper for all users currently logged in to the system. This option is useful within a Terminal Services environment or when BGInfo runs periodically on a system that more than one person uses.

- `/log`: BGInfo will write errors to the specified log file instead of generating a warning dialog box. Logging is useful for tracking errors occurring, particularly when you have the task scheduler run BGInfo.

- `/rtf`: BGInfo writes output text to an RTF file, including all formatting and color information.

Here is an example of running BGInfo from the PowerShell console:

```
PS C:\foo> Bginfo.exe \\WIN11\C$\foo\BGINFO.BGI /timer:00 /nolicprompt
PS C:\foo> |
```

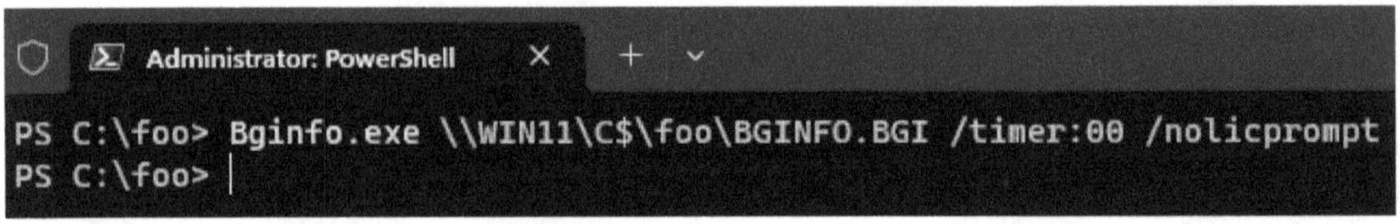

Figure 5.16 – Running BGInfo from the PowerShell console

This command runs `BgInfo.exe` and uses a saved configuration file (`\\win11\c$\BGINFO.BGI`). By specifying a timeout value of 0, BGInfo sets the background screen and exits. The `/nolicprompt` also eliminates the licensing popup that Sysinternals tools display the first time you run them.

Introducing the Sysinternals PsTools suite

PsTools is a part of the Sysinternals suite. These tools provide another method to perform remote system administration using the command line (via `cmd.exe` or PowerShell). Sysinternals first released these tools before PowerShell came onto the scene. While the remoting capabilities within PowerShell should satisfy the needs of most IT professionals, these tools provide additional assistance to the IT pro.

Installation is as simple as downloading the PsTools suite from `https://packt.link/tr896` and extracting the ZIP file. Execute them in an elevated console window. Setting them as part of your system path can be handy, but this can also be an attack vector, so carefully consider the impact before configuring.

The tools are Win32 console applications that perform different system tasks. You can use these both on local and remote systems. The PsTools suite contains the following tools:

- `PsExec`: Execute processes on remote hosts
- `PsFile`: Show files opened remotely
- `PsGetSid`: Display the **security identifier (SID)** of a computer or user object
- `PsInfo`: List information about a host
- `PsPing`: Measure network performance
- `PsKill`: Kill processes by either name or **process ID (PID)**
- `PsList`: List detailed information about processes on a host
- `PsLoggedOn`: See who's logged on locally and also via sharing
- `PsLogList`: Dump event logs
- `PsPasswd`: Change account passwords

- `PsService`: View and control services

- `PsShutdown`: Shut down and optionally reboot a host

- `PsSuspend`: Suspend processes

These are great tools that help IT pros. Some security professionals may want to block these tools to prevent abuse and/or provide attack vectors to others. You could use AppLocker to disallow the execution of these tools on given systems. For more details on AppLocker, see `https://packt.link/GC5NF`.

Some tools in the PsTools suite require network access to target computers and have services running. Port `445` is used by several of the tools. Several tools require the remote server to run the Remote Registry service, the server service up and running along with the IPC service. As a result, some of these commands do not work within PowerShell.

You may find that using PowerShell and PowerShell cmdlets for remote administration results in better security. You can configure PowerShell to perform logging (as noted in *Chapter 2, Introduction to PowerShell*), which enables you to know precisely who ran which commands and when (and what the output was). You can employ **Just Enough Administration** (**JEA**) to lock down remote servers further. With JEA, you create custom remoting endpoints only certain users can access. You also define the commands users can use and which parameters (and parameter values) they can specify.

JEA can take some time to configure (that is, determining which administrator needs which modules, and so on), but it provides superior security and auditability. For example, you could define a custom endpoint that allows any DNS admin's security group member to use the `Stop-Service` and `Start-Service` commands within suitably configured custom remoting sessions. JEA might then enable the user to run the `Stop-Service` command and specify the `-Displayname` property but only allow the user to specify the DNS service on `DNS1.contoso.com`.

For more details on JEA, see `https://packt.link/ABqi7`. And if you want to see JEA in action, see the script at `https://packt.link/DBHKD`.

Installing PsTools

PsTools uses a familiar syntax if you are already used to using a command line in Windows. Open an elevated console, using either `cmd.exe`, PowerShell, or Windows Terminal. Unless you installed the tools using the Microsoft Store, you should navigate to the folder where you expanded the executables.

Using your elevated console, type `psexec.exe` and hit *Enter*. The first time you run any of these tools, they prompt you to read and accept the EULA of the Sysinternals suite of software:

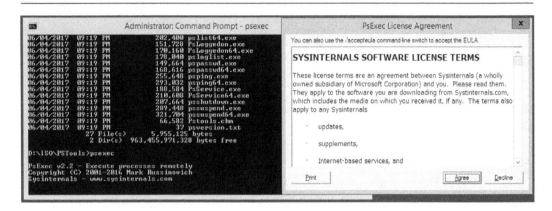

Figure 5.17 – Sysinternals licensing dialog

To avoid hitting the Sysinternals EULA notice on every system you use it on, use the `-accepteula` switch. You also see this with Process Explorer, **Process Monitor** (**ProcMon**), and most of the Sysinternals tools. Check out this command:

```
C:\Sysinternals> psexec \\win11 -u Reskit\JerryG -p SugarMagn0lia -c
-f procmon
```

It gives you the following output:

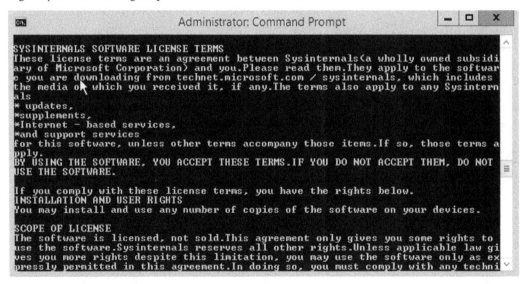

Figure 5.18 – Viewing license terms for Sysinternals tools

And it ends with the following:

```
This is the first run of this program. You must accept EULA to
continue.
Use -accepteula to accept EULA.
procmon.exe exited on win81 with error code 1.
```

Using the switch gives you better results (note that `-c` is to copy the binary locally, and `-f` is to specify the file).

Sometimes, it is necessary to chain commands. For example, `PsInfo.exe` requires the Remote Registry service to be enabled. With Windows 11, Microsoft set this service to disabled by default. So, if we wanted to use PsInfo to host `win11`, we'd need something such as this:

```
C:\Sysinternals\psservice.exe \\win11 setconfig remoteregistry auto &&
C:\pstools\psinfo \\win81

PsInfo v1.78 - Local and remote system information viewer
Copyright (C) 2001-2016 Mark Russinovich
Sysinternals - www.sysinternals.com

System information for \\win11:
Uptime:                        6 days 13 hours 23 minutes 55 seconds
Kernel version:                Windows 10 Enterprise, Multiprocessor Free
Product type:                  Professional
Product version:               6.3
Service pack:                  0
Kernel build number:           25375
Registered organization:
Registered owner:              tfl
IE version:                    9.0000
System root:                   C:\WINDOWS
Processors:                    64
Processor speed:               2.5 GHz
Processor type:                Intel(R) Xeon(R) Gold 6142 CPU @
Physical memory:               130572 MB
Video driver:                  NVIDIA Quadro K2200
```

> **Note:**
> In this example, PsTools sets the Remote Registry service to automatic, but in reality, this service is set to automatic (triggered) and only starts when needed.

Let's take a look at how you can use some of these great tools.

Using PsTools

The PsTools suite is a set of 26 console applications. You can download the tools directly from the internet. Alternatively, the tools come inside the Sysinternals suite you obtain via the Microsoft Store, as shown earlier in this chapter.

Some of these tools are improvements on the console applications shipped with Windows. For example, `PsPing.exe` is an improvement over the `ping.exe` application shipped natively within Windows. The PsTools suite's developers originally developed these tools for use within `cmd.exe`. But for the most part, they work well in PowerShell. Let's now look at some of these tools.

PsExec

This tool remotely executes a process on a single or multiple computers. The command redirects the input and output of any console application so that they appear to be running locally. A typical use case is opening a remote computer console window. From there, you can enter other commands such as `ipconfig /all` and see the output from that remote computer. This command is simpler than logging on remotely to get this information.

The basic syntax to use looks like this:

```
psexec \\computername [options] program [arguments]
```

To run `ipconfig` on a remote computer, you can run the following command:

```
psexec \\win11 ipconfig /all
```

If you need to use alternative credentials, you can run this command:

```
psexec \\win11 ipconfig /all -u "domain\userid" -p password
```

Many switches are available with this tool, and it is a good idea to review them on the help page to understand their potential uses (`https://packt.link/vIkw3`). Note that the `-u` and `-p` switches are common in many of the tools within the PsTools suite.

PsFile

The main purpose of this tool is to query a computer to list the files currently open on remote systems. To see files opened on a remote system, use the following syntax:

```
psfile \\computername C:\filepath
```

If you need access to a file currently locked by a process, you can close it using the `-c` parameter, but proceed cautiously, as this parameter may cause data loss. Be very careful!

PsGetSid

Windows SIDs allow unique identities for computers, groups, users, and other system objects. While the visible name of an object may change, the SID always remains unique for the object's life. Each computer has a machine SID created by the Windows installation process. If you join the computer to an Active Directory domain, it also has a **relative ID (RID)** applied to create a unique SID in the domain. This way, you can query information based on a specific identity without relying on names that may be changed or duplicated. For more details on SIDs in Windows, see this article: `https://packt.link/50h71`. The PsGetSid tool provides a simple way of translating from the name to a SID or vice versa. Use the following command syntax to display the SID of a given account name:

```
C:\Sysinternals> PsGetsid.exe reskit\jerryg
PsGetSid v1.46 - Translates SIDs to names and vice versa
Copyright (C) 1999-2023 Mark Russinovich
Sysinternals - www.sysinternals.com
SID for reskit\jerryg:
S-1-5-21-140053678-4069492383-922506915-1111
```

> **Information**
>
> For more details on SIDs, see `https://packt.link/zzvfC`.

PsInfo

This tool gathers key information about a system, including the operating system, product version, number and type of processors, physical memory, and video driver. This information can be useful to check the system compatibility before installing software upgrades or to query all computers and find those that don't meet your minimum requirements or have a specific application or hotfix installed. See the following example:

```
psinfo \\dc2
PsInfo v1.79 - Local and remote system information viewer
Copyright (C) 2001-2023 Mark Russinovich
Sysinternals - www.sysinternals.com
System information for \\dc2:
Uptime:                    6 days 7 hours 2 minutes 37 seconds
Kernel version:            Windows Server 2022 Datacenter,
Multiprocessor Free
Product type:              Standard Edition (Domain Controller)
Product version:           6.3
Service pack:              0
Kernel build number:       20348
Registered organization:   Reskit.Org
Registered owner:          Bobby Weir
```

```
IE version:              9.0000
System root:             C:\WINDOWS
Processors:              12
Processor speed:         2.5 GHz
Processor type:          Intel Xeon(R) Gold 6142 CPU
Physical memory:         2986 MB
Video driver:            Microsoft Hyper-V Video
```

You can use the following parameters to gather further information and control the report format and contents:

- -c to send the results to a CSV file
- -d to gather disk information
- -h to show all installed hotfixes
- -s to list all software applications installed
- -t to specify a different delimiter character

PsList

When you need to perform a detailed inspection of a remote computer, it is useful to understand which processes are currently running on the system.

By default, this tool displays some basic information. You can use the following parameters to have PsList show additional details about one or more processes. Due to the details included, you may want to run these against specific processes or export them to a CSV file to compare all processes:

- -d to display details about the threads for each process
- -m to display memory information about each process
- -x to display CPU, thread, and memory information about each process

You can combine these parameters to show detailed information about processes. The following example shows PsList being used to list all details about a specific application on a remote computer:

```
PS C:\Sysinternals> pslist.exe \\dc2 -x -e dwm
PsList v1.41 - Process information lister
Copyright (C) 2000-2023 Mark Russinovich
Sysinternals - www.sysinternals.com
Process and thread information for dc2:
Name                     Pid      VM       WS      Priv Priv
Pk    Faults    NonP Page
dwm                      544
2151955656    41148    23512    26460    306135    33    425
  Tid Pri    Cswtch                State      User Time    Kernel
```

```
Time    Elapsed Time
528  15       492    Wait:UserReq  0:00:00.000   0:00:00.031
151:06:08.022
4116 15      6591    Wait:UserReq  0:00:00.171   0:00:00.203
151:05:53.819
4124 15     56890    Wait:UserReq  0:00:00.015   0:00:00.000
151:05:53.816
4144 15      9572    Wait:UserReq  0:00:00.000   0:00:00.000
151:05:53.809
4152 16      1580    Wait:UserReq  0:00:00.046   0:00:00.062
151:05:53.807
4156 16       172    Wait:Queue    0:00:00.000   0:00:00.000
151:05:53.807
4160 16        46    Wait:UserReq  0:00:00.000   0:00:00.000
151:05:53.807
4200 15         2    Wait:UserReq  0:00:00.000   0:00:00.000
151:05:53.755
4204 13         7    Wait:Queue    0:00:00.000   0:00:00.000
151:05:53.714
4216 15        97    Wait:UserReq  0:00:00.000   0:00:00.031
151:05:53.675
4276 13     21035    Wait:Queue    0:00:00.015   0:00:00.015
151:05:53.643
4300 15        10  Wait:DelayExec  0:00:00.000   0:00:00.000
151:05:53.636
4308 13        38    Wait:UserReq  0:00:00.000   0:00:00.000
151:05:53.629
4352 15        61  Wait:LpcReceive 0:00:00.000   0:00:00.000
151:05:53.573
4364 15     58705    Wait:Queue    0:00:00.015   0:00:00.000
151:05:53.547
2464 15         1    Wait:Queue    0:00:00.000   0:00:00.000
7:04:52.914
1028 13         7    Wait:Queue    0:00:00.000   0:00:00.000
0:03:06.243
6044 13         1    Wait:Queue    0:00:00.000   0:00:00.000
0:00:04.797
```

You can also use the -t switch to show the processes in a tree view and the parent/child relationship. You *cannot* use the -t switch with the -d, -m, and -x switches.

Of course, there are PowerShell commands that get this same information that may be easier to use.

PsKill

Once you have used PsList to identify processes running on a system, you can use PsKill to terminate the process. PsKill is a powerful tool that ends the process immediately. Immediately killing a process means it probably won't have the opportunity to shut down cleanly, which can result in data loss or system instability. You can specify either the name or PID of the specific process or the whole process tree (parent and all child processes). The following command connects to the machine specified and terminates the process with a given PID:

```
pskill \\computername PID
```

To terminate the process tree, specify the -t switch.

PsService

This tool lets you remotely view and control services and drivers. The basic controls include starting, stopping, restarting, and pausing a service. You can query services and drivers for specific information, such as dependent services, security configurations, and load order (on boot). You can also configure a service or driver to change details, such as the account name under which the service runs, or change the load order so that it occurs earlier or later upon the next boot. *-Service PowerShell cmdlets may be more useful than this command.

A unique ability of this service is the search feature. You can search your network for all instances of a given service using the find parameter. For example, to find all computers running the DNS service, use the following command:

```
psservice find "dns server" all
```

Note that you need the Computer Browser service running on all systems in your domain or workgroup. This service uses the SMB 1.0 protocol, which is a security risk. SMB 1.0 is disabled by default in Windows 11. In most cases, the effective result is that the psservice command is not usable on Windows 11.

For a better understanding of the SMB 1.0 protocol and its weaknesses, see this article: https://packt.link/6Nts0.

PsLoggedon

When investigating client computer activity, it is important to understand whether whoever has connected to the computer is connected locally or remotely. To show a list of all users logged on to a specific computer, use this command:

```
PS C:\Sysinternals> \PsLoggedon.exe \\dc2
PsLoggedon v1.35 - See who's logged on
Copyright (C) 2000-2016 Mark Russinovich
Sysinternals - www.sysinternals.com
```

```
Users logged on locally:
     27/05/2023 09:24:07          RESKIT\Administrator
Users logged on via resource shares:
     02/06/2023 16:33:42          RESKIT\Administrator
```

This ends our look at some of the key components of the PsTools suite.

Custom code repository

With all these tools and flexibility at the hands of an administrator or, more likely, a team of administrators in an enterprise environment, there exists a great risk of duplication of effort as each administrator sees a problem and whips up a PowerShell solution, for example. Having a GitHub repository of scripts and tools makes sense so that you can maintain versioning of frequently used scripts as they are iterated on and improved. You can navigate to `http://packt.link/EDLC6` and create a custom repository for use by your admin team. You can point users in your organization to the GitHub repository as the source for key admin tools.

Also, if you are using PowerShell, you can create an internal module repository to enable your users/administrators to use `Find-Module` and `Install-Module` to install your own custom PowerShell modules.

You can create a simple repository using a file share on a remote system. The script at `https://packt.link/S4QCg` shows you how to set up and access a custom PowerShell repository.

Summary

To summarize this chapter, we learned about a few different options to manage your enterprise environment remotely. The key to understanding automation and administration is *to do, but do carefully*. We cannot overstate the risk of a badly written script executing in an environment with domain administrator credentials.

In the next chapter, you'll learn about device management in the enterprise and how to properly control device inventory, controls, and some configuration items.

6
Device Management

You have learned about remote administration and jump server configuration for troubleshooting, deployment, and general work-use scenarios in the previous chapters. In this chapter, we'll look at the new **mobile device management (MDM)** capabilities of Windows 10 and 11, discuss caveats of the Windows 10/11 **Group Policy Object (GPO)** processing, and have a deeper look at patching and servicing, including the deployment solutions of the needed quality and feature updates such as **Windows Update for Business (WUfB)**, **Windows Server Update Services (WSUS)**, **Microsoft Endpoint Configuration Manager (MECM)** (aka **System Center Configuration Manager**, or **SCCM**), and third-party solutions.

In this chapter, the following topics will be covered:

- Evolving business needs
- MDM
- Changes to GPOs in Windows 10/11
- Servicing and patching
- Update deployment solutions

Evolving business needs

According to Forrester Research, mobility is the new *normal*. Information workers will erase the boundary between enterprise and consumer technologies, and therefore mobility is certainly a defining vector in the evolution of the new business world. 56% of information workers send their first email before getting to the office, and 73% send their last email after leaving the office. 52% of information workers are using 3 or more devices for work.

Business needs are evolving with the new *Industry 4.0*, from employees working Monday to Friday, 9 to 5 toward a 24/7 blur of work and personal activity; from computers on a **local area network** (**LAN**) corporate network toward multiple devices, anytime, anywhere; and from on-premises applications and file hosting toward **Software-as-a-Service** (**SaaS**) applications and cloud-based file hosting.

With the onset of the pandemic in 2020, the world of work has changed for good. Working from home or remotely, a variety of video meetings, and changed working hours are just some of the effects and the *new normal*. Not only the infrastructure and the IT must adapt to it, but also the operating system. Windows 11 22H2 has some improvements to support the user in this situation better.

And we also need to emphasize something that Microsoft calls *Modern IT*. Times are changing quickly, and technical capabilities are advancing at a much faster rate than just a few years ago. So, to respond to this, you need a faster and more constant method of deployment to consistently deliver the best experience and most up-to-date technology to your users. The times of a user using old devices with an old OS and still being happy and working effectively are gone.

Additionally, COVID-19 has fundamentally changed how we approach work. With many organizations now having a significant portion of staff working remotely—and as things are looking, this is going to be the long-term reality—the old model of how companies support a *mobile* workforce is not exactly holding up well.

The old way of handing out corporate hardware doesn't work anymore. It's no longer just about issuing a laptop—the world and our requirements have changed. As we push closer to 3+ years of full- or part-time home working with no end in sight, the old model for what is considered *mobile worker* support on the hardware front is starting to show some serious gaps. Organizations have tried many things to cut down on the cost of maintaining employee workstations over the years—including moving whole classes of workers to Windows Terminal clients or other virtual desktops.

More about the improvements in *Chapter 10, Windows 11 21H2 and 22H2 Changes versus Win 10*. Also, old-school methods of managing computers need to evolve without increasing complexity over value. Windows 10/11 has enabled MDM without the need to install an extra agent. The following figure details the MDM capabilities of Windows 10/11:

MDM in Windows 10

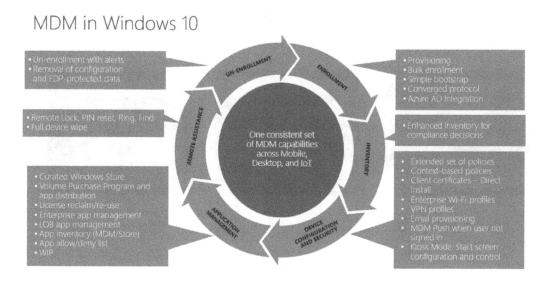

Figure 6.1 – MDM capabilities of Windows 10/11

In Windows 10, the MDM agent is already built in and usable with first-party (Intune/MECM) and third-party solutions. Windows 11 has the same MDM integration included. MDM management of Windows 10 and 11 is currently identical. MDM policies can also be created/applied by the Windows Configuration Designer (see *Chapter 3, Configuration and Customization*) or with a script and the integrated **Windows Management Instrumentation** (**WMI**) bridge. MDM policies can be used in domain-joined, Azure AD-joined, AD/Azure AD-hybrid-joined, and Azure AD-account-added scenarios. MDM can be used as a lightweight GPO replacement for computers joined only to Azure AD and mobile solutions such as Intune, AirWatch, or MobileIron.

The following figure shows the management options for Windows 10/11:

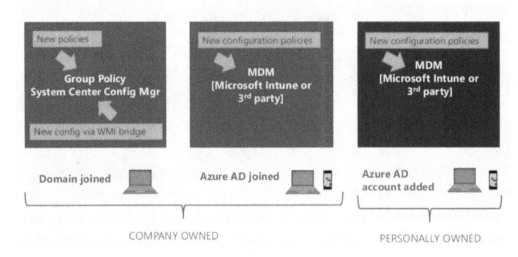

Figure 6.2 – Windows 10/11 management options

As all available configurations in Windows 10/11 can no longer be covered by GPOs alone (for example, **Windows Information Protection** (**WIP**) or Health Attestation/Provable Device Health), even without using Azure, you will be forced to use MDM or suitable scripts in conjunction with the WMI bridge or a Windows 10/11-compatible configuration solution such as Microsoft MECM, LANDesk, and HEAT.

As there are new MDM configuration settings with each new version of Windows 10/11, the configuration solution you use also needs to be upgraded to keep pace.

MDM

When discussing MDM, we need to look back in time to understand its origin and some of its limitations. Back in June 2002, the non-profit organization **Open Mobile Alliance** (**OMA**) was formed. OMA was merged with the **Internet Protocol Smart Objects** (**IPSO**) Alliance and rebranded to OMA SpecWorks (for more information, visit `https://packt.link/LPlLG`). The **OMA Device Management** (**OMA DM**) specification was originally designed for the management of mobile devices such as mobile phones, tablets, and **personal digital assistants** (**PDAs**). It was intended to provision and configure devices and enable software updates and fault management. There is a fixed set of OMA-DM protocol commands all vendors support. Currently, Windows 11 22H2 and higher supports MDM protocol version 14.0 (for more information, visit `https://packt.link/raIgs`).

MDM configuration objects are stored in a so-called **OMA Uniform Resource Identifier** (**OMA URI**) (for more information, visit `https://packt.link/4VuZ4`). You will need this OMA URI to add custom policies to your MDM solution if the setting is not available out of the box. You can think of the use of such a custom URI as similar to writing your own custom ADMX templates. As custom ADMX files need to write a supported registry key, the OMA URI needs to modify a supported resource identifier with a **configuration service provider** (**CSP**) capable of interpreting and applying the URI. Custom URIs can be added to Intune easily. Select **Windows Custom Policy** and fill out the **Add or edit OMA-URI Setting** box. Here is an example of the dialog box:

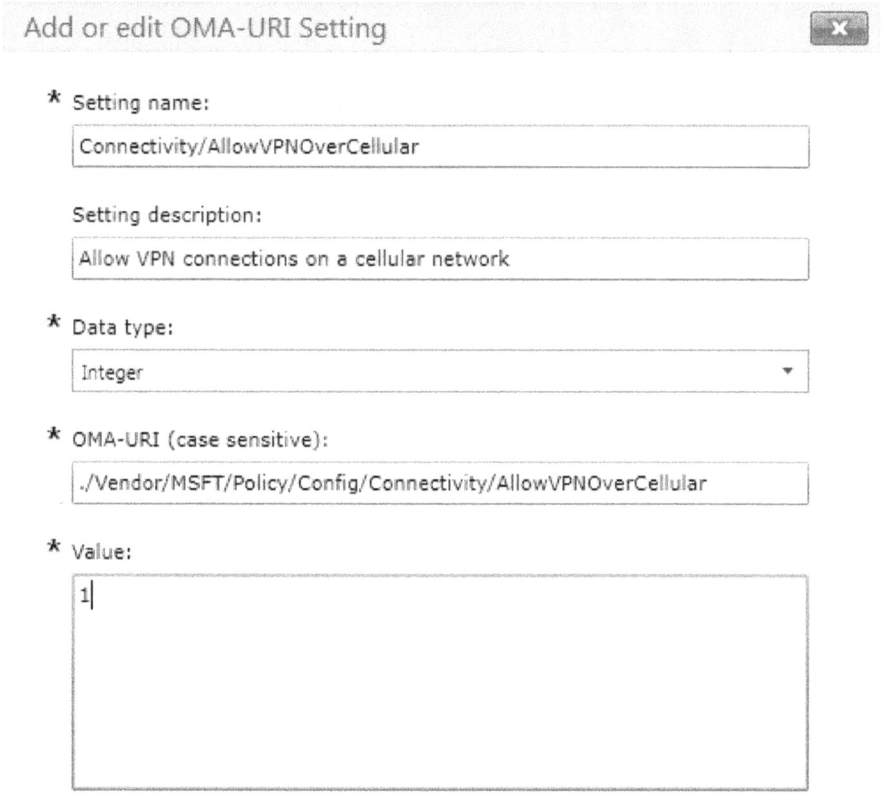

Add or edit OMA-URI Setting

* Setting name:

Connectivity/AllowVPNOverCellular

Setting description:

Allow VPN connections on a cellular network

* Data type:

Integer

* OMA-URI (case sensitive):

./Vendor/MSFT/Policy/Config/Connectivity/AllowVPNOverCellular

* Value:

1

Figure 6.3 – Defining custom OMA-URI settings

A list of all available CSPs in Windows 10/11 and their respective OMA URIs can be found at `https://packt.link/Av0oX`.

Every new release of Windows 10/11 brings new capabilities to MDM CSPs. An always-updated *TechNet* article can be found at `https://packt.link/IJ6Sv`.

New supported CSPs are also documented for each version at `https://packt.link/VIpkw`.

The following diagram shows as an example the BitLocker CSP in tree format. As you can see from `./Device/Vendor/MSFT`, it is a URI only applicable to Microsoft products:

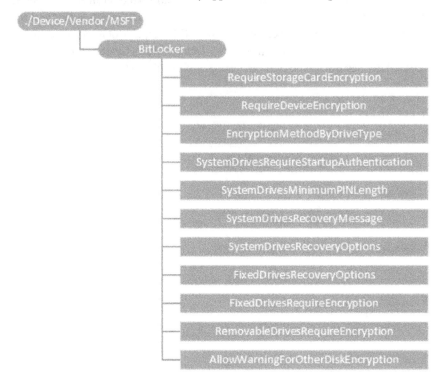

Figure 6.4 – BitLocker CSP tree

When configuring this CSP, a **Synchronization Markup Language** (**SyncML**) XML is generated and transmitted. Here is a (partial) sample of such a SyncML BitLocker XML:

```
<SyncML xmlns="SYNCML:SYNCML1.2">
  <SyncBody>

    <!-- Phone only policy -->
    <Replace>
      <CmdID>$CmdID$</CmdID>
      <Item>
        <Target>
          <LocURI>./Device/Vendor/MSFT/BitLocker/RequireStorageCardEncryption</LocURI>
        </Target>
        <Meta>
          <Format xmlns="syncml:metinf">int</Format>
        </Meta>
        <Data>1</Data>
      </Item>
    </Replace>

    <Replace>
      <CmdID>$CmdID$</CmdID>
      <Item>
        <Target>
          <LocURI>./Device/Vendor/MSFT/BitLocker/RequireDeviceEncryption</LocURI>
        </Target>
        <Meta>
          <Format xmlns="syncml:metinf">int</Format>
        </Meta>
        <Data>1</Data>
      </Item>
    </Replace>

    <!-- All of the following policies are only supported on desktop SKU -->
    <Replace>
      <CmdID>$CmdID$</CmdID>
      <Item>
        <Target>
          <LocURI>./Device/Vendor/MSFT/BitLocker/EncryptionMethodByDriveType</LocURI>
        </Target>
        <Data>
          &lt;enabled/&gt;
          &lt;data id=&quot;EncryptionMethodwithXtsOsDropDown_Name&quot; value=&quot;4&qu(
          &lt;data id=&quot;EncryptionMethodWithXtsFdvDropDown_Name&quot; value=&quot;7&qu
          &lt;data id=&quot;EncryptionMethodWithXtsRdvDropDown_Name&quot; value=&quot;4&qu
        </Data>
      </Item>
    </Replace>
```

Figure 6.5 – SyncML XML view of the BitLocker settings

As OMA DMs and OMA URIs originate from MDM, the design of these URIs favors integer values for their settings, which is quite alright for the OS but a bit uncomfortable for human readability. Therefore, you will need the corresponding CSP pages for translation very often. Here is an example of possible BitLocker encryption type values:

```
 <enabled/><data id="EncryptionMethodWithXtsOsDropDown_Name" value="xx"/><data id="EncryptionMethodWithXtsFdvDro
pDown_Name" value="xx"/><data id="EncryptionMethodWithXtsRdvDropDown_Name" value="xx"/>
```

EncryptionMethodWithXtsOsDropDown_Name = Select the encryption method for operating system drives

EncryptionMethodWithXtsFdvDropDown_Name = Select the encryption method for fixed data drives.

EncryptionMethodWithXtsRdvDropDown_Name = Select the encryption method for removable data drives.

The possible values for 'xx' are:

- 3 = AES-CBC 128
- 4 = AES-CBC 256
- 6 = XTS-AES 128
- 7 = XTS-AES 256

Figure 6.6 – Example of BitLocker encryption type values

Another thing to note when displaying MDM settings on your client is that there is currently no comparable tool to GPRESULT.exe built into Windows 10/11. You can get all applied settings by reading the registry, but you will just see a long list of values and not the originator of the value. Use the following PowerShell command line for reading the registry:

```
get-item 'HKLM:\Software\Microsoft\PolicyManager\current\device\*'
```

A resultant set of policies, such as a log file, can be exported in the system settings. Go to **Settings | Accounts | Access work or school | Export your management log files**:

Figure 6.7 – Work or school account view in Windows 11 settings

Unfortunately, this file is in plain XML style, which is hard to read. Since Windows 10 1903 and newer, an XML-to-HTML converter is now built in. Also, the log collection was further improved; an external tool is no longer needed. When downloading the diagnostic report, a ZIP file is created. The structure and content of the file are described here: https://packt.link/dNgYL.

MDM policies are applied on a fixed schedule. When joining/enrolling a Windows PC to your MDM solution, it will check for new policies every 3 minutes for 15 minutes, then every 15 minutes for 2 hours, and then run at its normal frequency of checking around every 8 hours (8 hours for Windows Mobile and Windows 10/11 desktop, 24 hours for (no longer supported) Windows RT). More about the policy refresh cycles can be read here: https://packt.link/qZC6o.

Last but not least, there is some added complexity due to the several origins of MDM settings in Windows 10/11. Besides the built-in MDM client, which can be connected to—for instance—Intune, AirWatch, or MobileIron, MDM settings can be induced by **Exchange ActiveSync** settings, the built-in EAS client, and the built-in WMI bridge used by MECM, or Windows PowerShell, as illustrated here:

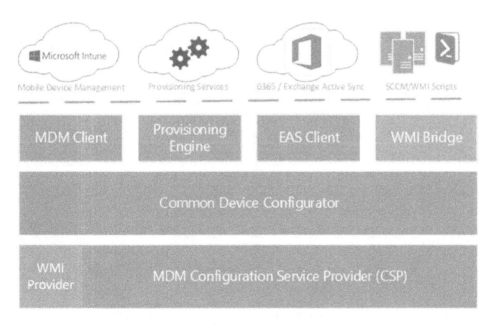

Figure 6.8 – WMI bridge diagram

Depending on whether the MDM setting is a security setting or a non-security setting, there are different override rules when applying them to a client. For security-related settings, the most restrictive setting will always win. For non-security settings, the base settings are Microsoft, **Original Equipment Manufacturer** (**OEM**), or enterprise-created PPKG packages. They will be overridden by EAS and MDM clients, and at the topmost priority is GPO (if the setting is also configured by GPO). This default behavior can be overridden; see the following link for more on this: https://packt.link/klB7h. Without using XML log troubleshooting, MDM settings are very hard and time-consuming.

On GitHub, there was a tool called the **MDM Migration Analysis Tool (MMAT)** in the `WindowsDevicesManagement` GitHub account to convert existing GPOs to new MDM settings.

Meanwhile, the ability to test existing GPOs for their convertibility to MDM has been built into Intune. As per the latest preview, the conversion itself is now also available in Intune via the **Migrate** option. As it is now built in, the MMAT tool on GitHub was retired and is no longer available.

The function can be found under **Microsoft Endpoint Manager admin center | Devices | Group Policy analytics (preview)**. To use it, one or more GPOs must be exported via backup in the Active Directory world. Then, log in to Intune on a computer with internet access and import the GPOs there with **Import**. The GPO names are also taken over. The analysis takes a certain amount of time, after which you can view the MDM support score as a percentage. The higher the value, the more settings can be converted:

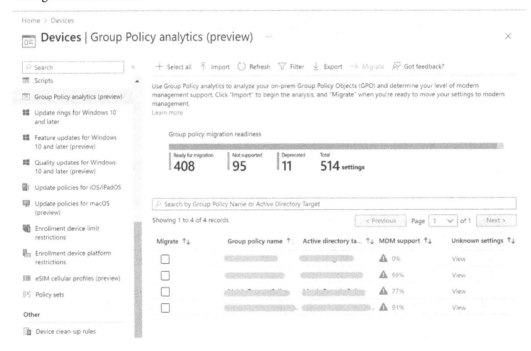

Figure 6.9 – Group Policy analytics (preview) screen in Intune with some GPO results

---

Important note about GPO-to-MDM conversion

Direct setting of registry keys via **Group Policy Preferences (GPPs)** is not supported by this conversion. So, it should be checked before whether the registry GPPs created earlier are now available as independent GPOs so that they can be automatically recognized by the system. It is also recommended to check the settings individually as to whether they are still necessary or useful for Azure AD instead of uploading all GPOs and converting them 1:1.

Changes to GPOs in Windows 10/11

Besides the major changes to MDM management, there are also changes to the GPO processing of Windows 10/11, which were first introduced with Windows 10 and will be covered now. These changes begin with GPOs only applicable to certain **stock-keeping units** (**SKUs**), known issues when upgrading your central policy definition store, and known issues when editing new GPOs, including GPPs with the old **Group Policy Management Console** (**GPMC**).

Enterprise- and Education-only GPOs

There have been policies that apply only to Windows 10/11, but for the first time ever in Windows history, there are also now GPOs, since Windows 10, that apply to certain SKUs only. Several GPOs for customizing Windows 10/11 apply only to Windows 10/11 Enterprise and Education SKUs. At the time of writing this book, the following GPOs have such a restriction:

- Configure Spotlight on the lock screen

- Turn off all Windows Spotlight features

- Turn off Microsoft consumer features (W10 only)

- Do not display the lock screen

- Do not require *Ctrl* + *Alt* + *Delete* combined with turning off app notifications on the lock screen

- Do not show Windows tips

- Force a specific default lock screen image

- Start layout and taskbar layout

- Turn off the store application

- Only display the private store within the Windows Store app

- Don't search the web or display web results

A full and updated list of group policies that apply only to Windows 10/11 Enterprise/Education editions can be found at `https://packt.link/jyHT7`.

There are expected to be more Enterprise- and Education-only GPOs in future releases of Windows 11, especially for more fine-grained UX control.

Known issues when upgrading the central policy store

Windows 10 and 11 ADMX files are currently not compatible as certain settings do not match or are only present in either Windows 11 or W10 ADMX templates. For these scenarios, consider using Central Store Override: `https://packt.link/qU47F`. Also, check out this blog post: `https://packt.link/5X3Fu`.

ADMX definition files are not only updated with every new release of Windows but sometimes also in between with **cumulative updates (CUs)**. You should always keep an eye on CU release notes and check your `PolicyDefinitions` folder for new entries from time to time.

With new ADMX files, not only are new settings available, but also old entries can be removed, renamed, or moved to a new category. When entries are removed and you've used them in your existing GPOs, you will see **Display names for some settings cannot be found. You might be able to resolve this issue by updating the .ADM files used by Group Policy Management** in the GPO report section:

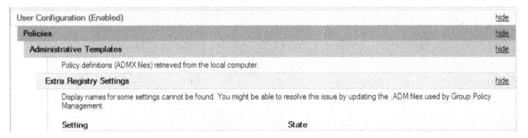

Figure 6.10 – Message in Microsoft Management Console (MMC) when ADMX files are missing

If old and new ADMX files with double definitions are in place, you will get a **Namespace 'abc' is already defined as the target namespace for another file in the store. File <xxxx>, line y, column z** error message, as seen here:

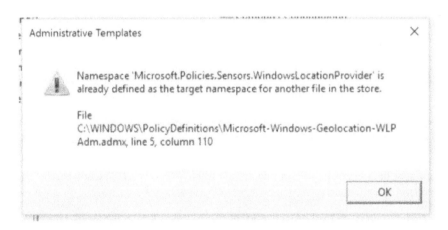

Figure 6.11 – Typical error message for ADMX double definition

In both cases, you need to carefully review all ADMX changes. To prevent such known issues while upgrading policy definitions/changes to ADMX files in different Windows versions, you should review `https://packt.link/N7mg4` and `https://packt.link/QBIZS`.

There is also a comprehensive GPO reference Excel spreadsheet describing all changes for Windows 11 22H2 at `https://packt.link/II67M`.

You should also update one client with **Remote Server Administration Tools** (**RSAT**) or a server with new ADMX definitions and then check every report of every existing GPO for these failures.

Also, review all your settings to see whether they are still supported under the new OS. A helpful entry point could be a Group Policy search; for more information, visit `https://packt.link/tSFAo` (Microsoft policies only) or `https://packt.link/F98SH` (Microsoft and third-party ADMX catalog).

It is recommended to always edit GPO/ADMX with the newest OS in use (also see the next issue).

Known issues with GPPs/GPMC

Normal GPO definitions are stored inside the ADMX files, and their translation is in the corresponding ADML file. When updating your `PolicyDefinitions` folder or your central policy definitions store on SYSVOL, you are able to create/define new GPO settings for the new OS. You could use older GPMC versions, and things would basically work.

GPPs are in total contrast to this behavior. They are hardcoded inside the GPMC. So, to get these new settings and filtering options, you need to use the newest RSAT tools or the newest GPMC on the server OS. Unfortunately, it is not only not seeing the new options when using an older GPMC, but you can seriously damage your GPO with GPP content when just opening it in an older GPMC.

When opening such a GPO with new GPP settings/item-level targeting in an outdated GPMC, the older GPMC does not recognize the new settings. So, at best, you might not be able to see all the options. For example, you would not find an option to filter on Windows 10 or Windows Server 2016 families, as seen here, in older GPMCs:

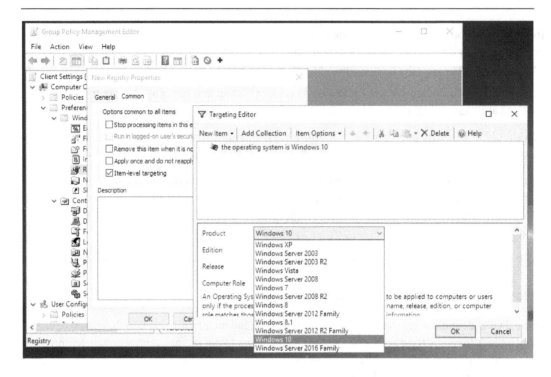

Figure 6.12 – GPMC view of target-level filtering

At worst, the older GPMC can interpret the new settings as **Corruption**. Corrupted settings are automatically repaired/removed. This repair attempt can trigger a **GPO was changed** event and therefore trigger an automatic save of the GPO. So, by just opening the GPO with GPP, you could accidentally remove settings without a notice/warning message. You would only notice an updated/higher revision number of your GPO.

So, always edit/administer new GPP settings only with the newest GPMC of Windows 11/Server 2022. To prevent such problems in multi-OS environments, when not all GPO/GPP editing systems can be updated at once, you should mark your new GPO/GPPs with—for example—_W11 and open such _W11 files only with the newest GPMC.

> **Tip**
> Always strive to use the newest GPMC of Windows 11 RSAT tools or the newest Windows Server (for example, 2022) as your management station to prevent problems when creating/updating the GPO with GPPs. Everything else would be suboptimal.

Servicing and patching

When we talk about changes to the way to service (or patch) Windows, it's important to first understand how things worked with Windows 7 and Windows 8.1. Each month, Microsoft released somewhere between 1 and 20 individual fixes for each one: some security updates, and some non-security updates. Most of these patches were **General Distribution Release (GDR)**, meaning available on **Windows Update (WU)**, WSUS, and Windows Update Catalog. Some patches were released under **Limited Distribution Release (LDR)** (also formerly known as **Quick Fix Engineering (QFE)**). LDR packages contain other fixes that have not undergone such extensive testing and resolve issues that only a fraction of the millions of Windows users might ever encounter. These LDR patches need to be downloaded on separate KB pages or sometimes requested from Microsoft services.

Most organizations deploy security fixes right away. But the non-security fixes sometimes aren't deployed at all, especially when talking about LDR non-security fixes. The result is that each organization ends up with its own unique Windows configuration, defined by the set of patches that it has installed.

Compare that to the configuration that Microsoft tests in its lab: fully patched PCs that have all the updates ever released installed. For each new update, Microsoft verifies that there are no adverse effects on these fully patched PCs.

But we've seen instances where these new updates cause issues on partially updated PCs (often with specific combinations of updates): Microsoft can't possibly test all these different possible combinations. And affected customers wonder why Microsoft didn't catch these simple issues when it did its testing. The following is one such example:

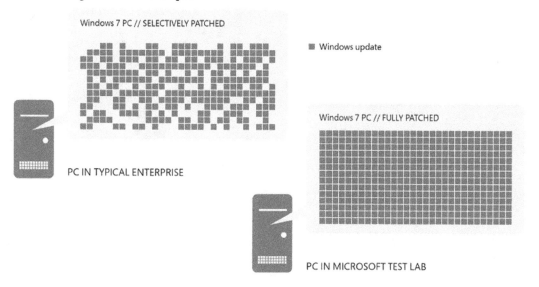

Figure 6.13 – Example of missing Windows 7 patches

For example, when speaking about Windows 7, there are more than 4,000 fixes since the release of SP1. About 600 of these patches are not widely spread. Now, try to calculate all possible combinations of patches if one or more of these 600 patches are missing.

Why CUs?

So, Microsoft decided that to improve the overall quality of Windows and to reduce the overall complexity of the patching process, it would rework the patching process altogether with Windows 10 (this changed servicing model is now also present in Windows 11). Let's explore these changes in more depth. Patches are now divided into **Quality Updates** and **Feature Updates**:

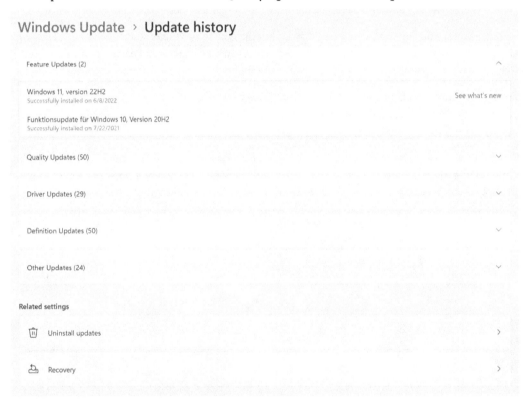

Figure 6.14 – Local Windows 11 update history

These so-called quality updates are a single monthly CU containing security fixes, reliability fixes, bug fixes, and so on. These CUs supersede the previous month's update. Normally, they contain no new features. Since 2017, there will be one mandatory CU on the second Patch Tuesday and possibly multiple CUs throughout the month with added non-security content. To stay on the secure side, you need to at a minimum deploy the second Patch Tuesday portion. The other patches can be optionally deployed on some or all systems. When Microsoft talks about Patchdays, you often hear the *B*, *C*, or *D* terminology. The easy explanation for it is the first Tuesday in the month gets the letter A, the second Tuesday gets B, and so on. The number in such announcements stands for the month. So, if Microsoft announces to plan to fix it with, for example, 10C, then it aims for the third Tuesday in October. Note that the schedule for preview updates changes in April: `https://packt.link/AaCEQ`.

Feature updates were done twice per year, each spring and fall, with new capabilities. This changed in 2022 when Microsoft decided to do only one feature update per year and set aside the second half of the year for the release. This change was made for Windows 11 and also for Windows 10 retroactively. Feature updates are technically simple deployments using in-place upgrades, driven by existing tools with built-in rollback capabilities. These are full upgrades with the exchange of the OS. It is also technically possible to feature updates as enablement packages (installed like a monthly update with just one reboot), but this limits the amount/kind of features that can be brought to the system. These enablement packages also need a certain base version to be applicable. The last four feature updates delivered for Windows 10 beginning with 20H2 were so-called enablement package feature updates. New features can be tested with Insider Preview. Please use this opportunity to familiarize yourself as soon as possible with the new features and configuration possibilities for the enterprise. Windows 10 22H2 will be the last Windows 10 version and will be supported till retirement in October 2025. So, you can fully concentrate on Windows 11 22H2 and newer.

Each quality update raises the version number of your Windows 10/11 release. You can see the quality update release build number as the last set of digits of `WINVER.exe` (for example, 730). The feature update raises the version itself (for instance, 22H2) and the build number's first set of digits (for example, 22621). The SKU of Windows 11 is in the fourth line (for example, Windows 11 Pro):

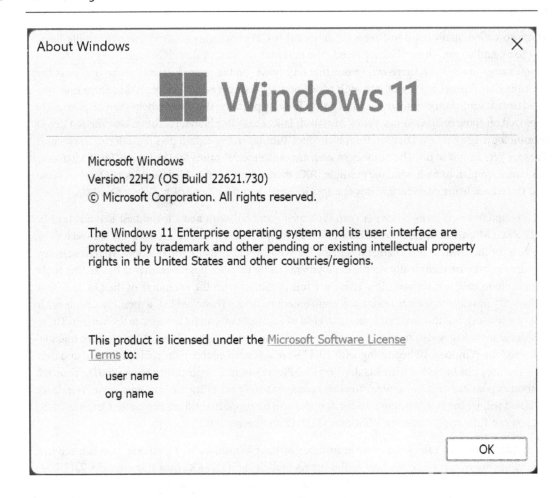

Figure 6.15 – Winver screenshot of Windows 11 22H2

A comprehensive and always updated list with the content of each CU can be found on the *Windows 11, version 22H2 update history* page at `https://packt.link/kKHXb`.

A Windows 11 release information page with a table containing all build numbers, release dates, and KB entries can be found at `https://packt.link/yLmSN` and for Windows 10 at `https://packt.link/HDlMF`.

As CUs are now all or nothing, it is no longer possible to exclude single patches if they break something in your environment. Due to always fully patched systems, there should be a reduced risk of incompatibilities, but it is still possible. So, you should pay special attention to the second Patch Tuesday CU and test/deploy it as fast as possible as it contains new security fixes. If there are any problems, report them to Microsoft right away so that it can fix them. Meanwhile, you can only uninstall/not deploy this CU and risk security flaws. When you uninstall a CU, your system automatically falls back to the last installed CU version. For non-security parts, you can test one to three extra CUs per month. Windows 11 22H2 starts to roll out additional features after **General Availability (GA)** in so-called **Moments**. The first released Moment was included in the Preview 10D CU and the regular 11B CU. The official name for Moments is **Continuous Innovation (CI)**. There was no way to block Moments' features when they were first released, and customers complained about it. So, with the release of the second Moments' features in early 2023, Microsoft is now shipping a subset of features deactivated by default if enterprise conditions are detected (domain, Azure AD, and so on). To get these features enabled, there is a central GPO/MDM policy to enable them. Microsoft states: "*Ship select features off by default and create a single policy (GP/MDM) that allows organizations to turn on these features. This will be done as a set, and not for individual features or individual releases.*" How you can configure the policy is explained here: `https://packt.link/ptipS`. Due to increased build numbers in Beta ring, we already know about more future Moment releases that are coming, but not the planned release dates.

If there is a known problem with the CU, single parts/functions of the CU can be deactivated by **Known Issue Rollback (KIR)**. More information about KIR can be found at `https://packt.link/CEuJy`.

These quality updates of early Windows 10 versions could grow very fast to sizes of 1 GB and more. With newer versions of Windows 10, the packaging was already improved, but with Windows 11 and its new internal CAB/PSF structure, the size was reduced even more. More details about **Unified Update Platform (UUP)** can be found here: `https://packt.link/VzLNn`. To know how Microsoft reduces the size of updates, you can check out the following link: `https://packt.link/BKYfg`.

To further reduce **wide area network (WAN)** traffic and/or workload on your on-premises servers, you need to configure **Delivery Optimization (DO)** (when using WU), BranchCache (when using WSUS), MECM peer delivery (when using MECM), the new **Microsoft Connected Cache (MCC)** (**DO In-Network Cache (DOINC)**), or the solution-specific peer delivery (when using a third-party solution).

> **How to distinguish between Windows 10 and Windows 11 programmatically**
>
> For compatibility reasons, Windows 11 states a lot of well-known registry paths to be *Windows 10*. As feature names *21H2* and *22H2* are used on both Windows versions, the only way to distinguish via script/program between Windows 10 and 11 is the build number. Windows 10 will stay in the range of 19xxx, Server 2022 is 20xxx, Windows 11 22H1 is 22000, Windows 11 22H2 is in the range of 2262x, where productive versions will be 22621 until further notice, and only Beta builds with activated Moments have build numbers 22622, 22623, and so on. More about these Beta builds in the *Windows 10/11 servicing* section later in this chapter.

Update deployment solutions

Updates can be deployed with different solutions. We will look at WU, WUfB, WSUS, management solutions such as MECM, and third-party solutions.

WU

The well-known WU relies on Microsoft cloud servers to patch and upgrade your systems. Upgrades are installed as they are released (subject to throttling in waves). To reduce the load on the servers and speed up delivery, optimization for **peer-to-peer** (**P2P**) distribution has been used since the first version of Windows 10. This is called DO and replaced **Background Intelligent Transfer Service** (**BITS**) as the main downloader in Windows. This update method is the only option for Windows 10/11 Home. Both Windows 10/11 Home and Windows 10 S SKUs do not support domain joining. Windows 10/11 Home has very limited MDM capabilities, while Windows 10 S can be managed and patched by an MDM solution. The options for P2P can be changed in the GUI under **Settings | Windows Update | Advanced options | Delivery Optimization**.

Windows 10 1709 introduced two new GUI entries, **Advanced options** and **Activity monitor**. With **Advanced options**, you can now specify exact limits for download and upload bandwidth and define a monthly upload limit, including an infographic showing how much is left. Bandwidth can be set between a minimum of 5% and a maximum of 100%, while the upload limit can be set between a minimum of 5 GB and a maximum of 500 GB. In newer versions of Windows 10/11, a download bandwidth in Mbit/s for the foreground and background can be defined by the GUI. Before 1709, all these settings were only available via GPO:

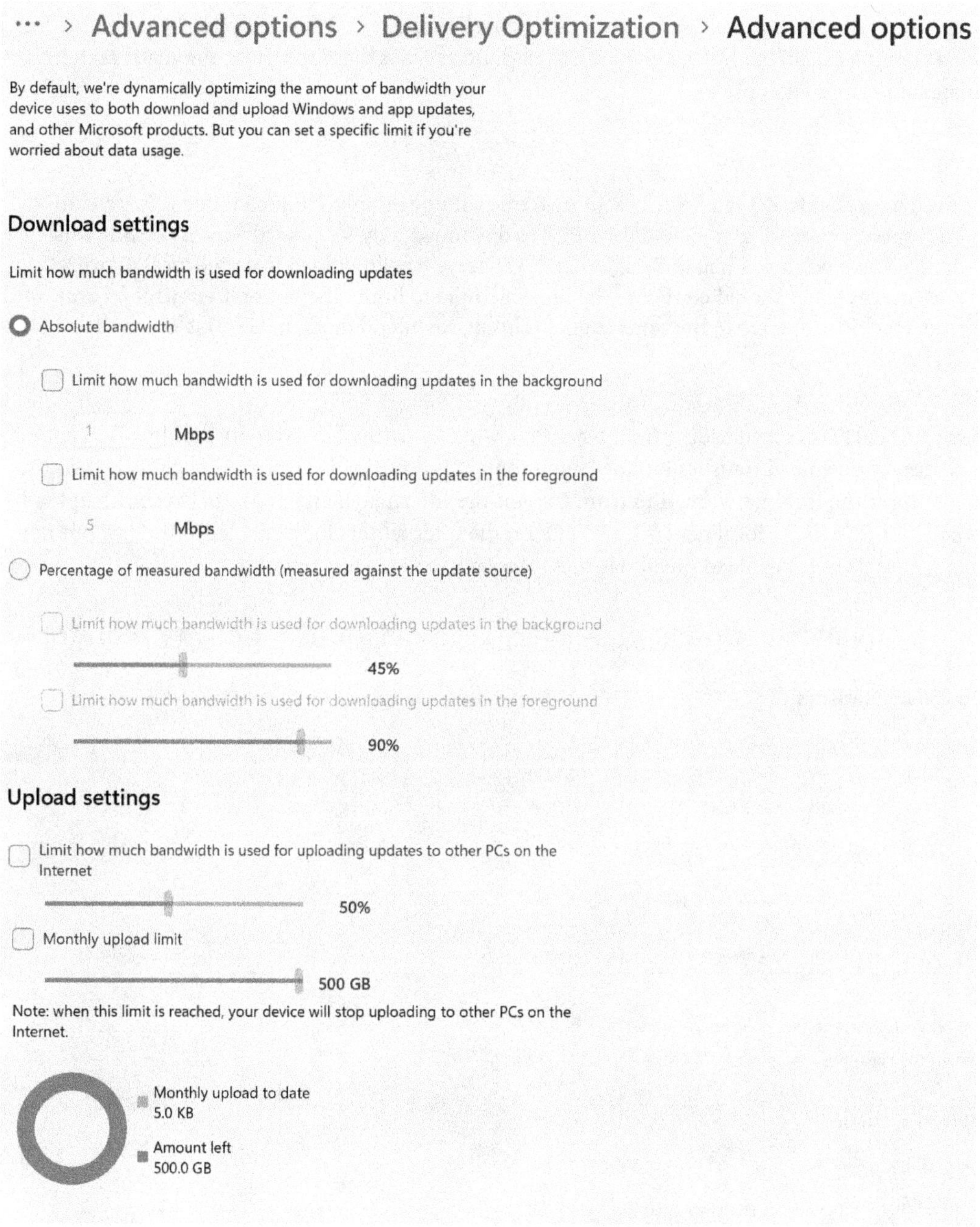

··· › **Advanced options** › **Delivery Optimization** › **Advanced options**

By default, we're dynamically optimizing the amount of bandwidth your device uses to both download and upload Windows and app updates, and other Microsoft products. But you can set a specific limit if you're worried about data usage.

Download settings

Limit how much bandwidth is used for downloading updates

⊙ Absolute bandwidth

    ☐ Limit how much bandwidth is used for downloading updates in the background

       1   **Mbps**

    ☐ Limit how much bandwidth is used for downloading updates in the foreground

       5   **Mbps**

○ Percentage of measured bandwidth (measured against the update source)

    ☐ Limit how much bandwidth is used for downloading updates in the background

    ─────────●─────────────   **45%**

    ☐ Limit how much bandwidth is used for downloading updates in the foreground

    ──────────────────●───   **90%**

Upload settings

☐ Limit how much bandwidth is used for uploading updates to other PCs on the Internet

─────●────────────────   **50%**

☐ Monthly upload limit

─────────────────────●   **500 GB**

Note: when this limit is reached, your device will stop uploading to other PCs on the Internet.

■ Monthly upload to date
5.0 KB

■ Amount left
500.0 GB

Figure 6.16 – DO Advanced options GUI

Foreground refers to the situation when a user actively triggers DO (for example, by searching for updates or updating in the Microsoft Store). Background means that the regular automatic search for updates/store apps takes place.

> **Tip**
>
> If you have an internet rate with a monthly traffic volume or only a limited traffic volume with high-speed bandwidth, you should set P2P to download only from local PCs. By setting this option, you block uploading to the internet. If you leave it at **download from the local network and internet**, you should configure the monthly upload limit. The option to download only from local PCs that are in the same subnet (Group download mode in GPO) is not available via the GUI.

To see the benefits of downloading from other PCs, you can use the new **Activity monitor** GUI entry. It will display download statistics for how much content is downloaded from Microsoft WU directly, from PCs on the local network, and from PCs on the internet (if enabled). It also shows upload statistics for PCs in the local network and PCs on the internet (again, if enabled). And last but not least, it shows some download speed statistics. The statistics are reset monthly:

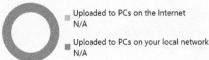

··· › Advanced options › Delivery Optimization › **Activity monitor**

Download Statistics

Since 10/1/2022

From Microsoft
100.00% (603.0 MB)

From Microsoft cache server
0.00% (N/A)

From PCs on your local network
0.00% (N/A)

From PCs on the Internet
0.00% (N/A)

Average download speed (user initiated): 27.4 Mbps

Average download speed (background): 14.4 Mbps

Upload Statistics

Since 10/1/2022

Uploaded to PCs on the Internet
N/A

Uploaded to PCs on your local network
N/A

Figure 6.17 – DO Activity monitor

There are multiple PowerShell cmdlets for analytics of DO, such as `Get-DeliveryOptimizationStatus`, `Get-DeliveryOptimizationPerfSnap`, and `Get-DeliveryOptimizationLogAnalysis`. More details about these cmdlets and their values can be found at `https://packt.link/DhBh7`.

If you are using a domain join-capable Windows 10 version such as Pro, for Workstation, Enterprise, or Education, you can define all these and even more fine-grained settings by GPO. You will find the relevant settings under **Computer Settings | Administrative Templates | Windows Components | Delivery Optimization**:

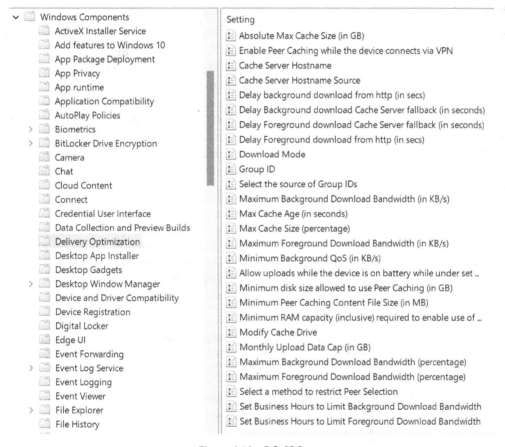

Figure 6.18 – DO GPOs

With Windows 10 1709, these settings were again extended. Now, you can define a DOINC server via GPO. DOINC Version 1 was later rebranded as MCC. By caching all DO queries, MCC can reduce WAN traffic significantly. DOINC/MCC v1 can only be used together with MECM and only deployed on **Distribution Points (DPs)**, which blocks the usability in branch offices without a DP/server. For more information about MCC v1, see `https://packt.link/NL6IR`.

At the time of writing this book, MCC v2, often referred to as MCC Standalone, is still in private preview. MCC v2 will leverage containers and will be able to run on Hyper-V, so it can be deployed on Server 2019/2022, Windows 10 1809 and newer, or Windows 11. This enables better scaling as it can be deployed on serverless branch offices as well as on dedicated server hardware in large enterprises. A special version for ISP is also planned. For more information about MCC v2, see `https://packt.link/MS62s`.

Please see the former *TechNet* aka `doc.microsoft.com` aka `learn.microsoft.com` documentation for further information about MCC v2 and its benefits as soon as information is released after GA in 2023.

Besides all these fine-tuning settings, the most important setting in this section will still remain in **Download** mode. With this GPO, you can disable the new DO completely (**Bypass**) and use the old BITS instead (see the next information box). You can limit DO to use only HTTP download without peering (**HTTP only (0)**), to use the internet and local PCs (**Internet (3)**), to use local PCs only when behind the same NAT (**LAN (1 – Default)**), to use local PCs only within the same AD site (if it exists) or the same domain (**Group (2)**). The **Bypass (100)** option is deprecated in Windows 11. It was replaced by the new **Simple (99)** mode, which disables the use of DO cloud services completely (for offline environments). When the DO cloud services cannot be accessed or are not available, or if the size of the content file is below 10 MB, the system will switch to **Simple** mode. In this mode, DO will ensure a dependable downloading process without utilizing P2P caching.

> **Please do not use Option Bypass (100)**
>
> This is deprecated and pending removal due to the confusion this option creates and its misuse, which can have an adverse impact on customers. For environments where you want to fully disable P2P, use the **Simple (99)** option instead. Note that multiple content scenarios either use DO or are in the process of integrating DO, which will not support BITS.

If you use MCC v1 with MECM, you should select the **Group** option so that MECM will generate unique GUIDs for each boundary to optimize peer and MCC download. In Windows 11, the **Local Peer Discovery** option was introduced to restrict peer discovery to the local network.

By selecting the **Simple** option, you use only HTTP but without contacting the DO cloud service. Most enterprise customers decide to use the **LAN** or **Group** option to prohibit upload:

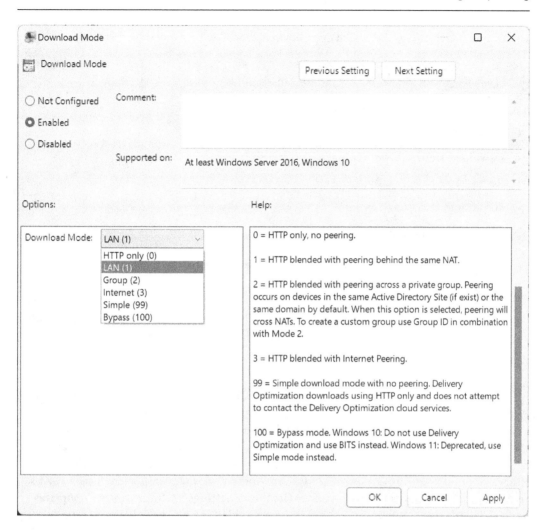

Figure 6.19 – DO mode

There are no general settings that will fit every environment. Good starting values for further tuning of DO are as follows:

- **Delay Foreground download from http**: 60 seconds
- **Delay Background download from http**: 180-300 seconds
- **Minimum Peer Caching Content File Size**: 1 MB
- **Download Mode**: **LAN** or **Group**
- **Restrict Peer Selection By**: Subnet
- **Minimum battery level**: 60%

When using MCC, additionally, consider the following values:

- **Delay Foreground download Cache Server fallback**: 60 seconds
- **Delay Background download Cache Server fallback**: 180–300 seconds

Depending on your business needs, you can additionally set business hours to limit foreground/background download bandwidth to, for example, 50% from 8 a.m. to 6 p.m.

WUfB

WUfB is often seen as an extra or new way of delivering updates to your clients, but it still uses WU for the content. It extends the classic WU with a set of configurations that enable the control of Windows 10/11 quality and feature update deployment. Updates and upgrades can be deferred, and preview builds can be managed. This helps small business users without their own on-premises patching infrastructure to build servicing rings and get a more fine-grained update experience. WUfB control settings are only available to Windows 10/11 Pro, for Workstation, Enterprise, and Education SKUs. The corresponding GPOs can be found under **Computer Settings | Administrative Templates | Windows Components | Windows Update | Windows Update for Business** (before 21H2):

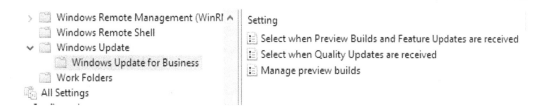

Figure 6.20 – WUfB GPOs (until 21H2)

Since 21H2 and newer, the corresponding GPOs can be found under **Manage updates offered from Windows Update**:

Figure 6.21 – WUfB GPOs (21H2 and newer)

The WUfB GPOs help to create update rings via GPO for monthly CUs (quality updates) and semi-annual servicing updates (feature updates). Update rings are explained in more detail in the servicing paragraph. If you are using WSUS, MECM, or third-party update solutions, you need to create target groups/collections in these solutions, as they will most likely ignore your WUfB GPO settings. By enabling and defining the **Select when Quality Updates are received** GPO, you can specify a delay between 0 and 30 days. You can also specify a date for temporarily pausing quality updates in the case of a known problem. When enabling the pause, it will remain in effect for 35 days or until you clear the start date field in the GPO. This GPO will have no effect if you set **Allow Telemetry** to **0 = Security only**:

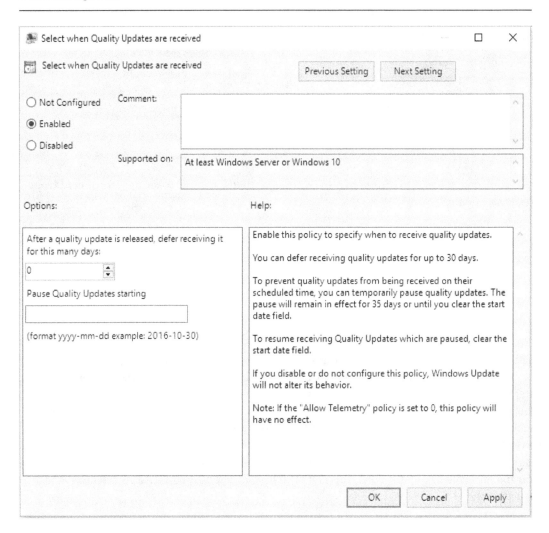

Figure 6.22 – Quality timing GPO

Another available WufB GPO is **Select when Feature Updates are received**. It was renamed in 1709 to **Select when Preview Builds and Feature Updates are received** and was capable of selecting timing and channel. The option to also select the channel was removed from the original **Select when Preview Builds and Feature Updates are received** GPO and moved to a new GPO. To select which channel to use for the Windows Insider program, a new GPO called **Manage preview builds** was created.

The former **Semi-Annual Channel (Targeted)** (**SAC Targeted**) (also formerly named **Current Branch**, or **CB**) and SAC (former **Current Branch for Business**, or **CBB**) channels were replaced by **General Availability Channel** (**GAC**). Besides that, there are multiple channels in the Windows Insider Program.

The Windows Insider Program has three preview build channels: **Dev Channel**, **Beta Channel**, and **Release Preview Channel**. **Dev Channel** is where Microsoft releases builds that are still in development and may contain bugs. This is not recommended for productive clients. **Beta Channel** is where Microsoft releases builds that are more stable than those in **Dev Channel** but still contain some bugs. **Release Preview Channel** is where Microsoft releases builds that are almost ready for release to the general public. If you only want to test quality updates, there is also now a fourth channel called **Release Preview of Quality Updates** where no new feature updates are distributed.

Upgrades to new Windows builds can be blocked centrally if certain incompatibilities are found. This is controlled by so-called **safeguard holds**.

The WU **Disable safeguard for Feature updates** policy is a policy that allows you to disable safeguard holds for feature updates on Windows 10 and Windows 11 devices.

Safeguard holds are used to ensure that devices are compatible with new feature updates before they are installed. For example, if a customer, a partner, or Microsoft internal validation finds an issue that would cause a severe effect (for example, rollback of the update, data loss, loss of connectivity, or loss of key functionality) and when a workaround isn't immediately available, then safeguard holds prevent a device with a known issue from being offered a new operating system version. Once the issue is verified as fixed, the update is made available to the device through WU.

To give enterprises the possibility to control (and if needed disable) these safeguard holds, a policy was introduced. This policy is available to WUfB devices running Windows 10, version 1809 or later that have installed the October 2020 security update and in Windows 11.

To disable the safeguard hold to upgrade Windows 10 to install version 22H2, you can use a GPO. You can also use the **Update/DisableWUfBSafeguards** CSP with MDM. After a device installs a new Windows version, the **Disable safeguards for Feature Updates** Group Policy will automatically revert to not configured (aka safeguards enabled) even if it was previously enabled (aka safeguards disabled).

More information about opting out of safeguard holds can be found here: `https://packt. link/iwdtW`

To build servicing rings via GPO, you can defer the servicing updates. When selecting GAC, former SAC, CB, or CBB, you can defer by up to 365 days (even though shorter ranges are recommended). As in the **Quality Updates** GPO, you can also specify a date for temporarily pausing servicing updates in the case of a known problem. When enabling the pause, it will remain in effect for 35 days or until you clear the start date field in the GPO. For deferring GAC, you need to set **Allow Telemetry** to **minimum 1 = Basic**.

Preview channel builds can only be deferred for up to 14 days or paused for up to 35 days. To defer any preview build, you need to set **Allow Telemetry** to **minimum 2 = Enhanced** and register your domain on `insider.windows.com`. More information about **Windows Insider Program for Business (WIPfBiz)** can be found at `https://packt.link/scajb` (*note that Enhanced does not apply to Windows 11*):

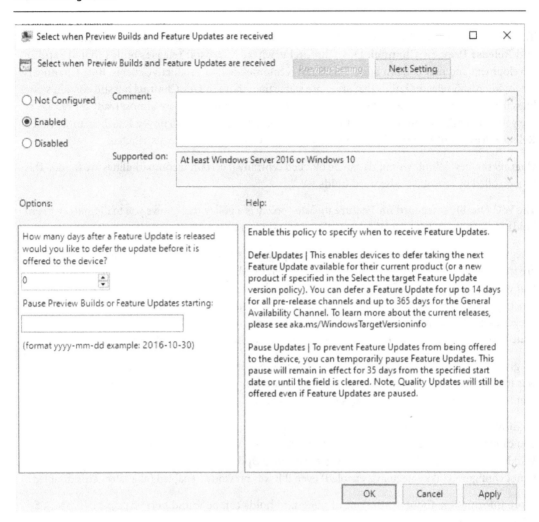

Figure 6.23 – Preview Builds timing GPO

Since 2020, there is a new option to define the target version (for example, 21H2, 22H2, and so on) and the Windows version (for example, Windows 10 or Windows 11). The GPO can be found under **Computer Configuration | Administrative Templates | Windows Components | Windows Update | Windows Update for Business:**

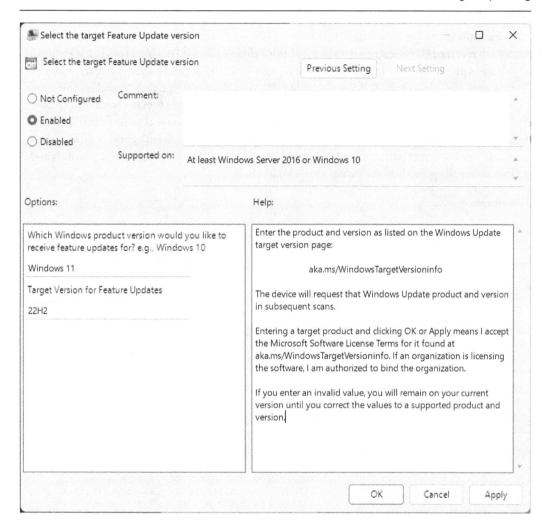

Figure 6.24 – Target Feature Update version GPO

Windows 10 1709 introduced a new WUfB GPO, **Manage preview builds**. You can select **Disable preview builds** to prevent installing previews on that device and prevent users from opting into the Windows Insider Program via the GUI. Selecting **Enable preview builds** will allow installing or opting into insider builds on this machine.

To automatically install preview builds, you additionally need to configure the **Select when Feature Updates are received** GPO described earlier. The third choice, labeled **Disable preview builds once next release is public**, will automatically halt the reception of insider builds when the current Insider Preview transitions into **Release to Manufacturing** (**RTM**) and becomes publicly available. This third option will currently not work for Developer and Beta builds. Only Release Preview works as expected with this option. From time to time, Beta and Release Preview overlap and will allow us to move from Beta to Release Preview. If you miss that time frame, you need to reinstall the client to leave the insider program. If you select that option, it will gracefully opt the device out of testing release previews and prevent accidentally going into the next preview build phase:

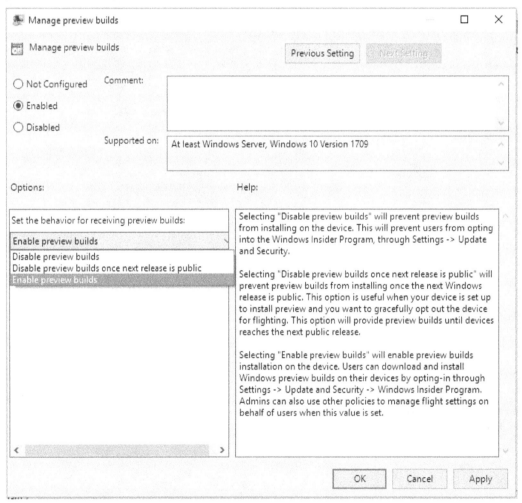

Figure 6.25 – Enable preview builds GPO

The mentioned GPOs only apply when using WU/WUfB and have no impact when using WSUS, which we'll describe shortly.

Windows Autopatch

Windows Autopatch is a new cloud service from Microsoft that automates Windows, Microsoft 365 Apps for Enterprise, Microsoft Edge, and Microsoft Teams updates to improve security and productivity across your organization. It helps minimize the involvement of your scarce IT resources in planning and deploying updates for Windows, Microsoft 365 apps, Microsoft Edge, or Microsoft Teams. It creates representative device groups in test rings and keeps them up to date.

Windows Autopatch helps you reduce the impact of updates on your organization by creating careful rollout sequences and communicating with you throughout the process. It also allows you to automatically pause, reset, and selectively perform updates to make your work easier and reduce risks. You can use Windows Autopatch with Microsoft Intune to deploy it on your devices. You can also use other ways to automate updates, such as using PowerShell scripts or third-party tools.

Windows Autopatch needs a Windows 10/11 E3 license or higher, Microsoft Intune, and a pure Azure AD or hybrid-join environment.

More information about Windows Autopatch can be found at `https://packt.link/zygxj`.

WSUS

WSUS is the first solution of all the update mechanisms mentioned earlier in the *Update delivery solutions* section to be on-premises. If configured to download content to your WSUS infrastructure, updates are distributed from your WSUS servers, which significantly reduces WAN traffic. Updates and upgrades are deployed when you approve them to your WSUS-defined target groups. You need to build your update and servicing rings via target groups inside WSUS.

> **Tip**
>
> At the time of writing this book, Windows 10/11 supports WSUS Server 2012/12R2 with KB3095113+KB3159706 and WSUS Server 2016/2019/2022. Due to better driver handling and for better support of future Windows 10/11 releases, we recommend you use at least WSUS Server 2016 or newer as soon as possible.

With the default setting of WSUS, it will download the full-size update packages and deploy them to the clients. This will rise very quickly to 1 GB per CU per client per month. To reduce the workload on your WSUS infrastructure, you should configure BranchCache to reduce bandwidth usage on the WSUS server.

There was a new option available at the end of March 2023 to download the UUP platform to your local on-premises infrastructure. This will help to reduce download sizes to the client and make all language packs and **Features on Demand** (**FODs**) available offline. Currently, it only works together with WSUS and ConfigMgr 2203 and later, and on the client side, it only supports Windows 11 22H2. More details about the new UUP on-premises are documented here: https://packt.link/U05IC.

Another option is to activate the **Download express installation files** option on WSUS, as shown here:

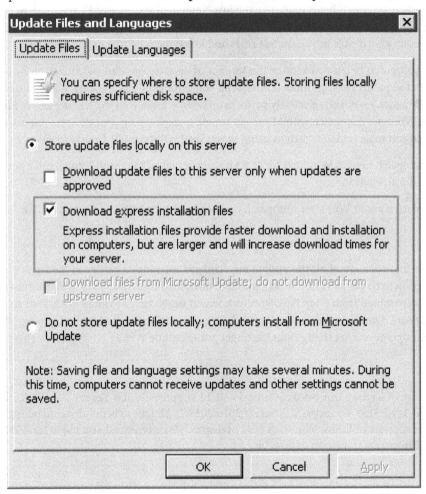

Figure 6.26 – (Dangerous) express installation files setting in the WSUS console

> **Tip**
>
> Be careful and double-check available storage before activating the **Download express installation files** option on WSUS. It will not only download Windows 10 express files but express files for all configured product classes on your version of WSUS. Depending on which products are selected, the required storage amount will be 4-12 (!) times higher than before. In particular, older products such as Office 2003/2007/2010 and Windows XP/Vista, if still configured on your version of WSUS, will increase storage more than other products. This can easily result in 2 TB+ files. There is currently no option for WSUS standalone configuration to select only Windows 10 for express files. When WSUS is controlled by MECM, there is a solution for downloading only Windows 10 express files, as described next.

WSUS is rather good at deploying monthly CUs/quality updates. However, there are several caveats and problems when deploying servicing/feature updates with WSUS. Some of the limitations and caveats are set out here:

- WSUS servicing media cannot be updated manually. You will deploy the RTM version until re-release by Microsoft some months later. Also, later on, you have no option to update servicing media after that.

- There is no option for a task sequence. Scripts and installations that need to be executed before or after an in-place upgrade are hard to target and a lot of hassle.

- Language packs and suitable configuration files to customize setup with parameters need to be placed before targeting a client for an update. There is no built-in check for file existence or similar. Again, a lot of fiddling and custom scripting is needed. (This is solved by the new UUP on-premises possibilities, but you will need W11 22H2 and WSUS and ConfigMgr 2203 or later.)

- To get dynamic updates working during WSUS feature updates, you can create a file at `%systemdrive%\Users\Default\AppData\Local\Microsoft\Windows\WSUS\SetupConfig.ini` to configure the `DynamicUpdate` parameter or point to a local folder with all language components. All parameters of the `setup.exe` command line can be specified in the INI file.

Michael Niehaus, former director of product marketing for Windows at Microsoft, explained in his public talks about Windows as a Service and Windows 10 four years ago that there are improvements planned for WSUS for future versions to avoid some of the mentioned problems. This will help small- and medium-sized business customers still using WSUS standalone. A first preview started for the newest Server 2022 together with MECM for Windows 11. Details of this are available at `https://packt.link/m2Rw1` and `https://packt.link/Ac7PD`.

MECM and third-party solutions

Using WU for servicing updates is not an option for business customers, especially at large-scale enterprises, due to missing task sequence functionality. Small- and medium-sized business customers using Pro and higher SKUs can use WUfB and WSUS for updating, but the missing task sequence will complicate the update.

You will run into a situation where you need to update a driver or software before being able to upgrade. Or, you may get into a situation where you need to do additional configuration steps and clean up after the in-place upgrade.

Five-plus years ago, when Autopilot was in the early stages and our day-to-day office work was different, my recommendation was that MECM and third-party solutions (such as LANDesk, HEAT, and many others) are the best solutions for serving updates and in-place upgrades. As with WSUS, quality and feature update content distributed from on-premises, such as configuration manager DPs, will significantly reduce the use of WAN bandwidth. Upgrades can be extended using a scripted task sequence, and you get extended software update capabilities in addition. Since then, a lot of things have changed – there's more home offices and decentralized working, a lot of improvements to Autopilot, and a new work style (and there are several more great features on the upcoming roadmap of Autopilot). This leads me to the revised recommendation that you should (if not already done) heavily invest in Intune/Autopilot. On-premises solutions are still supported, but we need to move on to the new deployment style. Autopilot has a lot of benefits, and the drawbacks will be overcome with every new release to Autopilot each quarter.

BranchCache and solution-specific peer delivery such as peer MECM delivery (client peer cache support for express installation files for Windows 10 and Office 365 available with MECM 1706 and newer versions) should be enabled to reduce bandwidth/workload on your servers. If you have not enabled BranchCache yet in ConfigMgr, have a look at DO. DO support in ConfigMgr has been greatly improved. There is also a new option in MECM 1702 and newer for downloading express packages for Windows 10 only (which also applies to Windows 11):

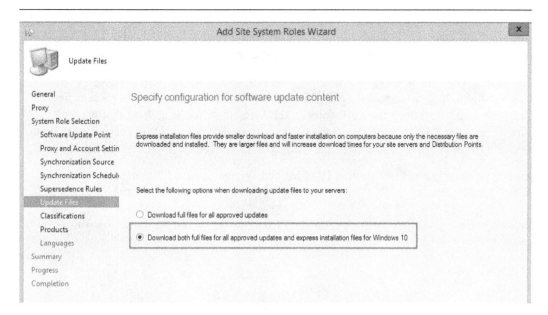

Figure 6.27 – Express file setting in MECM software update point (SUP) console

When selecting this option, it will only download express files for Windows 10 and 11, and therefore the needed storage amount on your version of WSUS will not excessively rise as in the WSUS standalone scenario described earlier.

For building updates and servicing rings, you need to use MECM or solution-specific techniques, such as MECM collections. MECM will ignore the GPO settings for quality updates and feature updates.

When performing traditional deployments that involve wipe and load (aka replacing the original image with a company image or in-place upgrades on Windows 10/11), it's important to use an updated version of the Windows 10/11 **Assessment and Deployment Kit (ADK)**. This updated ADK should include the latest Windows **Preinstallation Environment (PE)** unless exceptions are granted by the product group. Please review the MECM supportable matrix with every new Windows 10/11 release and every Windows release for the minimum version of MECM needed. An always updated matrix can be found at https://packt.link/4xRl3 and https://packt.link/eYKgA.

If you are using Autopilot, you can decide which OS comes preinstalled with a minimal image on your new devices (for example, Windows 10 22H2 or Windows 11 22H2). There are no dependencies on any ADK in that scenario.

At the time of writing this book, the MECM team is still planning three releases a year, and the supportable matrix does not include Windows 11 22H2 yet, but should be updated shortly after RTM/GA of Windows 11 22H2. If no serious issues or blockers are detected until the release of 1709, it can be expected to get backward compatibility for MECM 2203 and be fully supported for MECM 2207.

Here is a current view of the Windows 10 to MECM version compatibility matrix (https://packt.link/kmdtP):

Windows 10 version	ConfigMgr 2111	ConfigMgr 2203	ConfigMgr 2207	ConfigMgr 2211	ConfigMgr 2303
22H2 (10.0.19045)	✕	✕	⊘	⊘	⊘
21H2 (10.0.19044)	⊘	⊘	⊘	⊘	⊘
Enterprise LTSC 2021 (10.0.19044)	⊘	⊘	⊘	⊘	⊘
21H1 (10.0.19043)	⊘	⊘	⊘	⊘	⊘
20H2 Note (10.0.19042)	⊘	⊘	⊘	⊘	⊘

Figure 6.28 – MECM support for Windows 10

There is an extra page describing the Windows 11 to MECM version compatibility matrix (https://packt.link/KjewV):

Windows 11 version	ConfigMgr 2111	ConfigMgr 2203	ConfigMgr 2207	ConfigMgr 2211	ConfigMgr 2303
22H2 (10.0.22621)	✕	✕	⊘	⊘	⊘
21H2 (10.0.22000)	⊘	⊘	⊘	⊘	⊘

Figure 6.29 – MECM support for Windows 11

And there is also an extra page for the Windows ADK to MECM version compatibility matrix (https://packt.link/2sgke):

Windows ADK version	ConfigMgr 2111	ConfigMgr 2203	ConfigMgr 2207	ConfigMgr 2211	ConfigMgr 2303
Windows 11 (10.1.22621.1)	✕	⊘	⊘	⊘	⊘
Windows 11 (10.1.22000)	⊘	⊘	⊘	⊘	⊘
Windows Server 2022 (10.1.20348)	⊘	⊘	⊘	⊘	⊘
Windows 10, version 2004 (10.1.19041)	⊘	⊘	⊘	⊘	⊘

Figure 6.30 – MECM support for ADK versions

This is what the symbols stand for:

Key
⊘ = Supported
✕ = Not supported

Figure 6.31 – Key explanation

Other third-party solutions such as LANDesk/Ivanti Endpoint Manager, HEAT, and many others have already announced updates to their deployment solutions at least one to two times a year to keep up with Windows 10 and 11 and fully support the deployment. For example, LANDesk will release yearly major releases called **LANDesk Management Suite 2021.1**. Minor updates will be named SU1, SU2, and so on. In former times, the second release in the year was named **.2**. The next major updated version in 2023 will be 2023.1, and so on. A support matrix for LANDesk and Windows 10/11 can be found at https://packt.link/ijy9I.

Please review the solution-relevant support matrices and also plan to update your deployment solution in a higher cadence than with former operating systems.

Windows 10/11 servicing

The pros and cons of wipe and load versus in-place upgrades were already discussed in *Chapter 1, Installation and Upgrading*. When you successfully jump on the Windows 10/11 train, you need to plan to upgrade to new versions of Windows 10/11, at the latest, within 24-36 months, but you should transition much earlier.

Every new Windows 11 release will be in a preview phase for several months. During this preview phase, there will be several hundred builds created to stabilize the platform and integrate new features. Some builds are internal only or distributed to enterprise customers in a special technology adoption program. Builds with serious errors during internal validation are also not published. The preview builds are released in three different public-facing rings, called **Developer**, **Beta**, and **Release Preview**. Other rings are internal or for special private previews on invitation.

The **Release Preview** ring enables you to test feature updates and optional updates several weeks in advance. The Release Preview ring can normally be left without reinstalling the OS. This ring has a low risk of breaking your computer.

The Beta ring will preview features that will be released *in the coming months*. Currently, the future Moments updates of Windows 11 22H2 are tested. Not all computers in the Beta ring will be enabled to get A/B tests. You can distinguish enabled versus not-enabled clients via the build number. Clients with all Moments features enabled will increase the build number. 22622 was the October Moment, 22623/22624, and so on are future Moments. If the client stays at 22621 in the Beta channel, it is used as a comparison device. Currently, it is unclear what is the officially supported way to leave the Beta channel without reinstallation of the OS. There are some workarounds published on the net for detecting and uninstalling some kind of enablement package for this build number increase, but there is no statement from the Microsoft side on whether this is a supported scenario.

Developer rings are builds with the newest work in progress. At the time of writing this book, build numbers were in the range of 25xxx. Features and functions seen in these builds are targeted for a future version of Windows, but can also be integrated into the Beta branch if needed. Normally, the Developer branch can only be left without reinstallation for a very short time frame when Developer and Beta are at the same build number range. Then, you can jump to the Beta ring (and later to Release Preview or leave completely) when the release of the next Windows version is near. Due to the nature of the very early builds in the Developer channel, there is a certain risk of features and functions not working or, in the worst case, completely failing clients.

Business customers should take a look at WIPfBiz at `https://packt.link/B3UJj`.

Consumers and business customers who do not want to register at WIPfBiz can join the normal insider program at `https://packt.link/WeSol` but possibly miss some business test scenario descriptions.

When participating in the Insider Preview program, you will get first-hand information about new or deprecated features. You will get new builds early and can test them with your software for compatibility at an early stage.

When participating in the preview, you can access the fast ring for a bleeding-edge experience with the newest release at a slightly higher risk of features not working. If you want to work on a more stable preview or want to save download bandwidth with fewer updates, you can join the slow ring with updates about every 2 weeks.

During this preview phase, your feedback is very valuable, and a lot of decisions and changes in Windows 10 were already triggered by customers participating. Don't miss the chance to actively shape the future of Windows 11 now.

If you do not participate in the Insider Preview phase, then the official RTM or GA of the new Windows 10/11 release should be the starting signal for validating the new Windows 10/11 release in your environment.

> **Tip**
> Please do not plan to extend the pilot phase much longer than 4 months as it will limit and significantly reduce the usage phase with no changes to the Windows 10/11 release. Give users a calm-down phase before jumping to the next version. If problems occur, it may be necessary to skip a release and go directly to a newer version. But this should be an exception and not your general planning.

There is no universal one-fits-all recommendation on how many rings each phase should contain. It depends on the number of clients, the different use cases of your clients (office PC, manufacturing PC, and so on), and how many issues were detected during piloting.

Most Enterprise customers started with 1 ring for Insider Preview, 2-3 rings for the Pilot phase, and 4-5 rings for Broad deployment, where the highest ring is for blocking issues. According to the preceding scale, here is a sample of building rings and their timing. You can use it as a basis and adapt it to your environmental needs. Don't use the week's recommendation as the absolute minimum; faster deployment times are possible and have been observed, especially when iterating this job for the second or third time.

The following table shows an examples of rings/servicing branches/timing:

Deployment ring	Servicing branch	Total weeks after GA
Preview	Windows Insider	Participate if possible
Ring 1 Pilot IT	Release Preview or General Availability Channel	GA + 0 weeks
Ring 2 Pilot business users	General Availability Channel	GA + 4 weeks
Ring 3 Broad IT	General Availability Channel	GA + 6 weeks
Ring 4 Broad business users	General Availability Channel	GA + 10 weeks
Ring 5 Broad business users #2	General Availability Channel	GA + 12-14 weeks as required by capacity or other constraints
Ring 5 Broad business users #3	General Availability Channel	GA + 14-18 weeks as required by capacity or other constraints
Ring 5 Broad business users #4	General Availability Channel	GA + 18-22 weeks as required by capacity or other constraints and blocking issues

Table 6.1 – Table with sample rings as a starting basis

> **Tip**
>
> With the full support of 36 months for each Windows 11 release and there being a new Windows 11 version about every 12 months, it is theoretically possible (and supported) to skip a Windows 11 release and directly jump to an even newer version of Windows 11. But this will significantly reduce the maximum pilot and deployment time to approximately 12 months before your old version runs out of support and will mean you will not receive any more security patches. If you want to jump, for example, from 21H2 directly to 23H2 and skip 22H2, you will get a pilot version of 23H2 when 21H2 has only support for 12 more months. So, you need to pilot, solve all issues, and broadly deploy on all clients in 12 months. As no fixed date for 23H2 has been announced yet, this may even be shorter than 12 months when 23H2 is released closer to the end of 2023. Please do not plan with this as a general basis for your rollout.

Summary

In this chapter, you learned about the new MDM capabilities and changes in GPO processing of Windows 10/11. In the servicing and update part, we discussed the different update delivery solutions and gave recommendations for building servicing rings to keep up with the fast Windows 10/11 release cadence. With the options presented, you should now be able to set all the necessary settings for successful and smooth patching, planning release rings, and optimizing patch traffic regardless of whether you are still on-prem or already modern with MDM/Azure AD. Please use, if possible, the presented possibilities for testing preview builds to be best prepared for new Windows versions.

In the next chapter, we will have a closer look at protecting enterprise data in **bring your own device (BYOD)** scenarios.

7

Accessing Enterprise Data in BYOD and CYOD Scenarios

This chapter discusses accessing enterprise data in **End User Computing** (**EUC**) device scenarios. The main objective of this chapter is to provide information and guidance on accessing corporate data on personally owned Windows 11 devices. To achieve that, the chapter will take us through EUC device models, key considerations, device choice options, ownership, and management responsibilities.

Note that securing EUC devices will be covered in *Chapter 8, Windows 11 Security*.

In this chapter, the following topics will be covered:

- What are the EUC device models?
- Protection and governance options
- Storage sync options
- Alternative EUC delivery options

What are the EUC device models?

In 2019/2020, the COVID-19 global pandemic changed so much about what we do and how we do it forever. We had our approach to **Bring Your Own Device** (**BYOD**) and **Choose Your Own Device** (**CYOD**) redefined.

Historically, IT catered only to a minority of an organization's workforce that operated outside of the company offices; now, in 2023, we are adjusting to a new normal of a *hybrid* workforce in most cases.

These changes have been reflected in our attitudes to EUC solutions, and how technology must adapt and evolve to meet an organization's productivity demands for its users.

The following sections will first introduce the concepts of the BYOD and CYOD EUC solutions; we will then look at key considerations such as device choice, ownership, and management responsibility, and compare the options.

The bring your own device model

In this section, we will introduce the concept of the BYOD EUC model. BYOD is an EUC solution approach in your endpoint strategy. It is the most commonly known EUC device solution and has existed almost since the first iPhone when a C-level exec wanted access to their work mail on their personal device. It became mainstream circa 2010, at about the time of the first iPad, as well as the increasing adoption of Android devices.

With this model, a personal device is used for both personal and work-related tasks. This is the same whether the user is an employee, an external or third-party contractor, or temporary staff. **Mobile Device Management** (**MDM**) and **Mobile Application Management** (**MAM**) solutions are combined to provide control and governance to access company data from these personally owned devices. We will look further at these topics in the *Protection and governance options* section later in this chapter.

There are many situations where you need to collaborate with individuals and other companies, but do not want to issue them one of your devices to gain access to your systems and information.

Many supply chain situations mean a corporate-owned or corporate-chosen device is unavailable to be procured in the required timescale; this has driven a new demand for the BYOD approach. We will look at the corporate-chosen approach in the following section.

The choose your own device model

In this section, we will introduce the concept of the CYOD EUC model. CYOD is another EUC solution approach in your endpoint strategy. This model allows users to choose their own devices based on minimum requirements, which are then managed using MDM and MAM solutions.

The user is still expected to follow the acceptable use policy but may also use the device for some personal use. This option provides a balance between end user mobility and information security. It works toward solving one of the biggest issues with any IT deployment, the engagement of the end user.

CYOD (and BYOD) devices have some challenges that must be addressed to apply the appropriate controls and ensure the user remains productive while the company data is kept secure.

User account administration is still one of the biggest risks to any IT system. Scenarios where a user can install software or make configuration changes by being granted local administrative access to their device may weaken the security and integrity of the system, either intentionally, accidentally, or through malicious intent.

From an operational perspective, costs can be reduced if the right policies are in place to manage and govern the life cycle of devices. Unfortunately, unlike BYOD, it does not relieve the burden of hardware costs.

Key considerations

Several key areas require review and consideration to assess requirements and risk factors. The following sections discuss device choice, ownership, and management responsibility considerations.

Device choice

Device choice may be restricted to a few standard options in a managed environment. However, the total cost of ownership for managed devices will be lower, with the simpler deployment of OS images, drivers, applications, and settings, and greater compatibility of accessories.

Compared to BYOD, the user has a choice from hundreds of options, depending on personal preference and budget. With this in mind, a minimum standard should be published to ensure users know what devices they should look for, such as the OS version, browser choice, and ability to support security features such as **BitLocker**, and hardware security such as a **Trusted Platform Module** (**TPM**).

Previously, it was considered a good practice to provide users with example devices that met these standards to ensure compatibility with company systems and services; this is not always possible with disrupted supply chains. An alternative is rather than providing a prescriptive list of example devices, we should publish guidelines for the type of devices that would be well suited for the required work tasks, explaining the benefits of the various choices and why the user might choose one device over another.

It is, however, also a good practice to publish a list of devices that are known to be incompatible and will cause the user problems if they try to use them for work tasks.

Ownership

One of the key cost-saving components in a BYOD strategy is transferring cost and, therefore, ownership to the user. Some companies will choose to provide a set monetary value they will contribute toward the cost of the device based on a 2-3-year lifespan. Simpler budgeting for the IT department can be enabled, removing the burden of depreciation and disposal at the end of the life of the device. It also allows the user to choose a device within the budget, or pay extra for a higher-spec device to suit their personal preferences and any accessories to improve productivity.

Either way, the device is theirs to keep at the end of the 2-3-year lifespan. This is the key difference between a personally owned BYOD device and a company-sponsored BYOD device. A user may be entitled to expect that their personally owned BYOD device is within their full control if they have paid for it, and therefore that it should not be managed by the company. In contrast, a company-sponsored BYOD device does not fully belong to the user until the end of any agreed service period to cover the cost of the device (consider what would happen if the user left the company within 12 months of receiving the allowance).

By contrast, CYOD devices are purchased and owned by the company. The company may choose to allow the user to keep, or buy back, the device at the end of its life cycle, but otherwise, it is handled in the same way as any other company asset.

Management responsibility

The user may choose the device to fit their personal requirements; they may purchase and even own it. However, they may not expect to maintain the configuration and security management requirements; this burden should still fall on the company.

BYOD and CYOD can be managed in different ways. The following table shows three options that can be considered for all device types:

Compare device management	MDM enrolled	MAM only	MDM + MAM
Level of risk	Low	Medium	Low
Management complexity	Medium	Low	High
Application deployment options	Company managed	User managed	Company managed
Conditional access and SSO	Enabled	Fewer options	Enabled

Table 7.1 – Compare device management options

Some users may want or need local admin rights to customize the device to their requirements, while others may expect their IT support to be able to manage and configure the device remotely on their behalf. Understanding and agreeing on who is responsible for the management of the device is key to ensuring that the appropriate level of security is applied.

These considerations then define the appropriate level of trust for each device. For example, if the user has local administrative rights to the device, they can modify the configuration, install software, and generally increase the risk profile. A user logging in to this device would therefore have a lower level of trust than on a device that is enrolled and managed by company policies and has the user's local admin rights removed.

Comparing options

Each option comes with its risks and benefits. One way to decide which is right for your company, or specific user groups within your company, is to list all the differences and compare them. You can use the following tables as a starting point for your assessment:

Compare device ownership	BYOD	CYOD
Range of device options available to the user	More	Fewer
Purchase costs and ownership	User	Company

Compare device ownership	BYOD	CYOD
Cost of damage and replacement	User	Company
Total cost of ownership	More	Less
Range of personal usage	High	Low
Management and control	Less	More

Table 7.2 – Comparing device ownership models

Now that we understand the difference between EUC models, that is, BYOD and CYOD, as well as the key considerations, it makes it easier to choose the model that is right for our company. Next, let's look at the protection and governance options for these EUC models.

Protection and governance options

Multiple options are available to provide the appropriate security and governance controls for BYOD scenarios. With the shift in security approaches from perimeter networks to end user devices, a defense-in-depth and layered approach should be taken; you can then identify which combination of options is required for your specific business requirements, technical capabilities, and end user scenarios.

The following topics will be covered in this section, specifically those related to BYOD and CYOD scenarios:

- Identity and access management
- Information protection
- Device configuration
- Application management:
  - Provisioning packages
  - Mobile application management

Identity and access management

In a scenario where a **corporate-owned device (COD)** is on the company network being **Active Directory (AD)** joined, then **identity and access management (IAM)** is generally controlled by AD and Group Policy. A device may even be **Azure AD joined** (or *hybrid joined*), where a solution may use **Microsoft Intune** policies.

However, in a BYOD scenario, the device may spend more time off the network, not registered or enrolled into any management solution; the end user may not want the restrictive policies applied in such a way for a personal-use device.

In the BYOD case where the user's device is not joined to the network with an AD join, they lose certain benefits, such as seamlessly signing on to applications from their device logon based on a common identity. Users will be prompted to enter their company credentials when they attempt to access resources, such as Microsoft 365. This can be called **same sign-on** but not **single sign-on**. We discuss IAM in more detail in *Chapter 8, Windows 11 Security*, covering **Windows Hello for Business**, **Microsoft Defender Credential Guard**, and **Microsoft Authenticator**.

Information protection

Microsoft's guidance outlines that information protection needs are now met using Microsoft Purview **Data Loss Prevention** (**DLP**). This solution has deep integration with **Microsoft Purview Information Protection**, offering to meet cross-platform and cross-cloud data protection needs. This robust information protection solution allows for the discovery, classification, and protection of sensitive information.

You can learn more about Microsoft Purview and DLP in the following Microsoft Learn article: `https://packt.link/S4rgF`.

The **Windows Information Protection** (**WIP**) solution, previously known as **Enterprise Data Protection** (**EDP**), is built into Windows 10 and provides isolation of company data and personal data going to the end of life.

In July 2022, Microsoft started deprecating WIP, and in December 2022, support was removed for the WIP without enrollment scenario in Microsoft Intune.

Support tip
End of support guidance for WIP: `https://packt.link/hA7Ve`

Device configuration

To ensure that a BYOD device meets the necessary security standards, you should ensure that it is enrolled in AD or Intune, or that the user enrolls in the MDM solution. These options will enable the central configuration of the security required. Some cloud services, such as Azure AD, can then use conditional access policies to ensure access is only granted to specific services if the device is compliant and domain joined. We will look at device security considerations such as TPM, BitLocker, and Microsoft Defender in *Chapter 8, Windows 11 Security*.

Application management

There are several options for deploying apps to Windows 11 devices. However, for those devices that are not part of the company network (that is, they are not managed by AD), there is a need to find alternative methods. If the device is enrolled in an MDM solution, then this can be used to advertise or force the installation of company applications. However, if the device is not enrolled, users can still gain access to company apps in one of the following ways.

Provisioning packages

Depending on your application deployment solution, it may be possible to provide your users with software packages they can install on their BYOD devices. These packages can be stored on a file share or cloud storage, or handed out via a USB memory stick.

Mobile application management

A solution such as Microsoft Intune can be used to create application control profiles. When the user installs the software and signs in using their corporate account, the MAM policy can enforce specific restrictions to ensure the application is used safely (such as a PIN or local disk encryption). If the device is not compliant, the application cannot be used, and any company data can be removed without impacting other applications and data on the device. All of this is possible without domain join or MDM enrollment.

Storage sync options

In this section, we will look at the available storage sync options of **OneDrive for Business** and **Work Folders**.

OneDrive for Business

This solution is a core part of the Microsoft 365 platform and provides a cloud storage and sharing solution. There are several options available to ensure that data is protected – for example, allowing users only to synchronize their OneDrive folders on authorized devices. If the device is not domain joined or compliant (for example, enrolled with Intune MDM), then the user will only be able to gain access to the content via a browser. Controls can also be set to control the ability for the user to share their content from OneDrive to internal or external third parties.

It is also possible to govern access based on device specifics, such as restricting access based on the IP address and support for modern authentication. If the MAM section is grayed out, then settings are being controlled by Microsoft Intune instead.

Work Folders

For those companies that are not ready to adopt a public cloud service yet, you could deploy SharePoint or the Work Folders feature that is part of Windows Server 2012 R2 and later. This feature enables secure access to files and folders via the internet. Device support includes Windows (11, 10, 8.1, and 7), Android, and iOS. Device policies can be configured to ensure devices meet specific requirements before connecting to files. All data can also be encrypted on the device, even if BitLocker has not been enabled.

You can learn more about work folders, how various Microsoft sync technologies are positioned, and when to use each in the following Microsoft Learn article: `https://packt.link/Edx31`

Alternative EUC delivery options

With the pandemic of 2020, we entered a new era of the hybrid workplace, and with it, the endpoint computing strategy is transforming at an ever-increasing rate of innovation.

Now more than ever, *work* is no longer somewhere *we go* as such; the PC, wherever that may be, is now *the office*.

Windows 365 Cloud PC and Azure Virtual Desktop

Traditionally, companies have implemented EUC solutions such as **Remote Desktop Service (RDS)/Virtual Desktop Infrastructure (VDI)** to provide users access to company apps and data. The last couple of years have seen innovation in this area.

Microsoft launched **Azure Virtual Desktop (AVD)**, previously called **Windows Virtual Desktop (WVD)**, a Microsoft desktop and app virtualization service that can utilize **multi-session** Windows 10 and Windows 11 images for desktops.

Layered on top of this desktop virtualization service hosted in Microsoft Azure, Microsoft provides **Windows 365**, also referred to as **Cloud PC**; it provides the ability to securely stream a personalized Windows experience to any BYOD/CYOD device. This fully Microsoft-managed PC solution has predictable costs due to it being licensed per user, per month, as opposed to the AVD consumption model.

The following figure outlines the EUC solution approaches for a hybrid workforce:

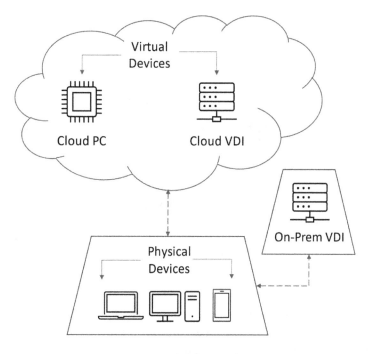

Figure 7.1 – EUC delivery solutions

There are several decisions a company must make to determine the best EUC approach for them; these include the following:

- **Pricing models**: Capital or operating expense via Azure consumption-based or per-user/per-month licensed models

- **IT skills/experience**: Physical endpoints, virtualized endpoints such as VDI/RDS, and Azure expertise

- **IT priorities**: Security, governance controls, operational efficiencies, modernizing existing VDI, data center migration, hardware refresh, license expiry, and so on

For the virtualized approach to EUC, the core differences are as follows:

- **Cloud PC – Windows 365**:

    - Optimized for simplicity

    - Predictable per-user pricing

    - Personalized persistent desktop for Windows 10 or Windows 11

    - Microsoft-managed service, complete from end to end

- **Cloud VDI – Azure Virtual Desktop**:

  - Optimized for flexibility

  - Flexible consumption-based pricing

  - Multi-session/non-persistent desktops for Windows 10, Windows 11, or Windows Server

  - Customer-managed solution; full control and responsibility over configuration, management, operations, and support

> **Note**
>
> AVD can be implemented as an on-premises VDI solution using Azure Stack HCI as the virtualization host platform. More information can be found in the following Microsoft Learn article: *Azure Virtual Desktop for Azure Stack HCI overview* (https://packt.link/GS3H2).

Enabling virtual private networks

If the device is managed and trusted, you can configure a **virtual private network** (**VPN**) to create a secure tunnel between the user's device and your company network or resources in a public cloud provider platform such as Microsoft Azure. This ensures that information cannot be intercepted across the network (such as in a public Wi-Fi hotspot); however, any data copied and stored on the device is still vulnerable to any local attacks against that device.

With the increase of a hybrid workforce, the VPN access method is often being evaluated to see whether alternative secure public internet access EUC solutions can replace it. Alternative solutions such as Cloud PC or Cloud VDI can be adopted, where the company apps and data are not stored on the local EUC device; they are now merely terminal devices.

Publishing applications via proxy

Another option to provide remote access to internal systems and data is to publish the internal system via proxy services. This service carries out the authentication and conditional access checks before granting access to the internal resource. This is an alternative to a VPN as it does not require local configuration or software installation on the user's device; however, HTTPS must be configured to enable traffic encryption.

End user behavior analytics

By monitoring activities in the logs, we can discover anomalous and suspicious user behavior and assess the potential risk of certain activities, such as the user's geographic location when they access the system. If it is not their usual location (such as Australia), then we can decide whether they should be blocked or at least prompted for an alternative authentication request (MFA). Microsoft offers this functionality as part of Microsoft Defender.

For further information on detecting and remediating impossible travel scenarios, please refer to the following Microsoft 365 Defender blog post: `https://packt.link/M7tjx`

Summary

In this chapter, we covered the BYOD and CYOD scenarios, key considerations for deciding which types of devices can be used by your users, and the risks and benefits of each option. Whether you enforce MDM to manage external devices or opt for a MAM-only option, there are plenty of choices for providing access and governance to resources. We also looked at alternative EUC solutions and storage sync options.

In the next chapter, we will explore the new hardware and software-based security options available in Windows 11.

Windows 11 Security

This chapter covers all aspects of Windows 11 security. While we have covered some aspects of security in some of the previous chapters of this book, we will look at them collectively and in more detail in this single security-focused chapter. If you are a security professional, this chapter is dedicated to your role and responsibilities in securing Windows 11 in a company.

The attacker's chain of events can be prevented and disrupted through a **zero-trust** and **defense-in-depth** approach. We need to put multiple obstacles in the attacker's way and increase their attack costs so that they will move on to launching an easier attack elsewhere that offers the least resistance. We will look at both of these approaches in subsequent sections of this chapter and understand why it is important to implement a good **security posture** to address these threats.

In this chapter, we'll learn about the following:

- Introducing security posture
- Zero trust
- Defense in depth
- Ensuring hardware security
- Ensuring that we operate system security
- Ensuring user identity security

Introducing security posture

A security strategy must start from an inward look at a company's current security position and **secure score**. A secure score is like a credit-rating score, but it looks at your positioning on the attack vulnerability scale of 1 to 10.

A **security posture** refers to an organization's current *threat-protection* and *threat-response capabilities*. This ensures that an organization has the ability for systems, data, and identities to be recoverable and operational should an attack be successful.

It is critical to understand that we cannot prevent or eliminate threats and attacks, and the fact is that an attacker only has to be successful once, while you must protect everything all the time.

A security posture's goal should be to reduce exposure to threats, shrinking attack surface areas and vectors while building resilience to attacks, as they cannot be eliminated.

A security strategy and security posture should use the guiding principles of **confidentiality**, **integrity**, and **availability**, also referred to as the **CIA triad** or **CIA triangle**. There is no perfect threat prevention or security solution; there will always be a trade-off, and the CIA model is a way to think about that. The CIA triangle is a common industry model used by security professionals; it is not a Microsoft model.

Let's look at these guiding principles in more detail here:

- **Confidentiality**: This is a requirement that sensitive data is kept protected and can only be accessed by those who should have access through the **principle of least privilege** (**POLP**). Confidentiality is the confidence that data cannot be accessed, read, or interpreted by anybody other than those intended to read and access it. This can be achieved by encrypting the data. The encryption keys also need to be made confidential and available to those who need access to the data.

- **Integrity**: This means that the data transferred is the same as the data received; the bytes sent are the same bytes received. Integrity is the confidence that the data has not been altered from its original form or tampered with. This can be achieved by hashing the data. Malware can threaten the integrity of systems and data.

- **Availability**: This means that data and systems are available to those that need them, including access to encryption keys, but in a secure and governed manner. Availability means a trade-off between the triangle's three sides and a balance between being locked down for security but accessible for operational needs and productivity. A DDoS attack will threaten the availability of systems, data, and encryption keys.

The *CIA triangle* model can be represented by the following diagram:

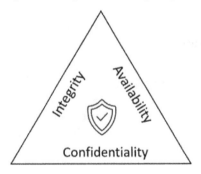

Figure 8.1 – Security posture CIA triangle

In this section, we looked at security posture. We will look at zero trust in the following section.

Zero trust

With many companies embracing a hybrid workforce, a new security model mindset is required more than ever. We need to adopt a holistic approach to security, a model that thinks beyond traditional network-perimeter-based security. The traditional firewalls and security-service-controlled network perimeters have vanished due to this hybrid workforce.

Zero trust, which uses the *never trust, always verify* approach, is not a service or solution but a wider security strategy and framework to be adopted. It ensures compliance and securing of access to the resources rather than the location or network it is on. We must not assume trust because of the device or resource's network or location. We can no longer assume trust based on identity or self-attestation.

The zero-trust framework is built upon the following foundational principles:

- **Assume breach**: From the start, we must adopt the mindset that there is a breach; it is all about damage limitation. As it is no longer about preventing your systems from being compromised, it's about limiting the impact *when* they are compromised. With measures in place, attackers are prevented from causing a maximum impact on systems and data. We can ruin their attempts to meet their goals and implement barriers such as protecting privileged roles, limiting lateral access, limiting data leaks and encrypted resources, and so on.

- **Verify explicitly**: We must *never trust* but *always verify*. Authentication and authorization should only succeed on verification from all data points and signals available, such as identity, app/service, location, time, device status/health, anomalous behaviors, and so on.

- **Use least-privilege access**: We must limit users' access by implementing *risk-based* adaptive policies, **just in time** (**JIT**), and **just enough access** (**JEA**).

The following are the zero-trust framework's six foundational elements:

- **Identities**: Users, services, devices; each represents an element to be compromised

- **Devices (endpoints)**: Represent an attack surface and threat vector for data flow

- **Applications**: Represent the consumer of the data flows

- **Infrastructure**: Represents an attack surface and threat vector, whether locally on-premises or remotely hosted by a cloud provider

- **Network**: Represents an attack surface and threat vector and should be segmented

- **Data**: Represents the data stored that is to be protected

In this section, we looked at the concept of zero trust. We will look at the concept of defense in depth in the following section.

Defense in depth

When considering securing Windows 11 in our enterprises, we should take a **defense-in-depth (DiD)** approach. This means we should not rely on a single security layer solution.

Adopting a DiD strategy allows an organization to adopt a strong security posture and helps ensure that all systems, data, and users are better protected from threats and compromise. A DiD strategy means no single layer of protection or security service is solely responsible for protecting resources. Still, you can slow down an attack path by implementing many different types of defense at individual layers. It may successfully breach one defensive layer but be halted by subsequent protection layers, preventing the protected resource from being exposed. The following figure shows that DiD as a concept is nothing new as a strategy; it can be considered the medieval castle concept of protecting resources:

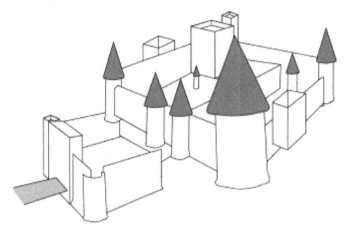

Figure 8.2 – Medieval castle defense approach

The medieval castle approach should be part of your strategy for building your resources in Azure. You define multiple layers that can be protected by different security services that are the most appropriate at each layer. As with our medieval castle analogy, each layer from the center to the outside to the center provides its own independent protection service, tailored to protect the characteristics of that layer best. The following diagram aids in visualizing the layers that make up a DiD strategy for a resource to be protected:

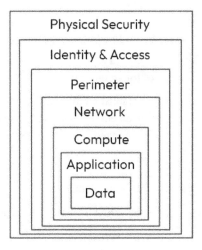

Figure 8.3 – DiD approach layers

No one-size-fits-all security service can protect all the layers. However, we must have security services at each layer that work in conjunction and complement the layers outside and inside of their layer. There must be a unified view so that telemetry and threat intelligence can be passed between each layer and enhance the protection at each layer. Microsoft enhances these capabilities by using **artificial intelligence (AI)**, threat intelligence, and analytics.

In this section, we looked at the concept of defense in depth. We will look at hardware security in the next section.

Ensuring hardware security

We need to secure our hardware through device protection. Hackers have historically easily compromised a device before it is booted by dropping in rootkit malware without it even being noticed. It remains undetected after the device starts.

Trusted Platform Module (TPM), the **Microsoft Pluton** security processor, **Hypervisor-Protected Code Integrity**, and **Windows Defender System Guard** are all measures that can be used to provide the integrity of the device and OS before it even starts up. We will look at these measures in detail in the following sections.

TPM

TPM is a hardware-based security measure that provides tampering protection and can provide **device health attestation**. At the heart of TPM is a secure *crypto-processor chip* used for actions such as *cryptographic key generation, storing,* and *use limitation*.

Device health attestation enables trust to be established for a managed device based on the hardware and software components under the control of an enterprise. An MDM solution can be configured for health attestation service queries, allowing a managed device to access a sensitive resource or denied.

The following device checks can be carried out:

- Is there support for **SecureBoot**, and is it enabled?
- Is there support for **Data Execution Prevention**, and is it enabled?
- Is there support for **BitLocker Drive Encryption**, and is it enabled?

> Note
>
> **TPM 1.2** is not supported for Windows 11. **TPM 2.0** provides *device health attestation* for Windows 10, Windows 11, Windows Server 2016, and Windows Server 2019.
>
> TPM 2.0 requires **UEFI firmware**. TPM 2.0 will not work as expected with a device with **legacy BIOS**.

The most common TPM functions include measurement of systems integrity, key creation, and use operations. The boot code (firmware and OS components) loaded during the systems boot process can be measured and recorded on TPM. This can provide an audit log of evidence and show that a TPM-based key was used to boot the system only when correct and integrity-intact software was used.

TPM-based keys can be made unavailable outside the system so that they cannot be copied and used without TPM; it is a good way to mitigate phishing attacks. TPM-based keys can also be configured to be required to use an authorization value. Authorization value guesses can be prevented by TPM's dictionary attack logic.

From Windows Server 2019 and Windows 10, version 1809, onward, Microsoft is no longer actively developing the TPM management console. It is recommended that the configuration of TM via the TPM management console, TPM.msc, is now avoided. In Windows 10 (and now Windows 11), TPM is automatically initialized and taken ownership of by the OS.

There are exceptions to this, relating to tasks such as resetting or cleanly installing a device. Further information on this can be found in this Microsoft Learn article: https://packt.link/Hsfkx.

Costs can also be reduced in an enterprise by automating the provisioning within TPM. Several settings via **group policy** can be utilized in enterprise scenarios; more information can be found on the TPM group policy settings in this Microsoft Learn article: https://packt.link/pZccD.

In this section, we looked at TPM. We will look at the Microsoft Pluton security processor in the following section.

Microsoft Pluton security processor

Microsoft Pluton is a security processor chip with proven TPM security technology used for **Azure Sphere** (IoT device security platform) and is also found in **Microsoft Xbox**.

Microsoft Pluton is a joint venture between Microsoft and silicon partners. It is built into the CPU as a secure cryptoprocessor. It provides the standard TPM functionality and other security functionality for Windows 11 devices beyond what is possible with the TPM 2.0 specification. It protects identities, credentials, encryption keys, and personal data. It is updated with firmware and OS features via Windows Update.

The following figure shows the basic architecture of the Microsoft Pluton system:

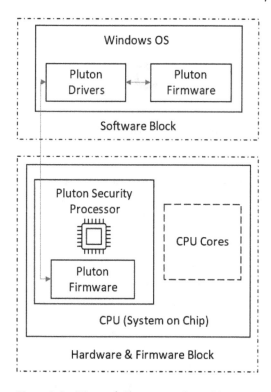

Figure 8.4 – Microsoft Pluton security architecture

As shown in *Figure 8.4*, Pluton consists of the following layers for its subsystem:

- **Hardware**: The Pluton security processor provides a trusted execution environment and delivers required cryptographic services for protecting resources such as data, keys, and so on.

- **Firmware**: The Microsoft-authorized Pluto firmware is stored on the motherboard's Flash storage and is loaded at system boot as part of the initialization of the hardware. A copy of the firmware is loaded into the OS during the Windows startup process; if available, the latest firmware is obtained from Windows Update. The firmware exposes the interfaces the OS and applications can use for Pluton interaction.

- **Software**: The Pluton subsystem provides OS drivers and applications with seamless usage of hardware capabilities to end users.

For a closer understanding of the firmware load flow, we can look at the following illustration:

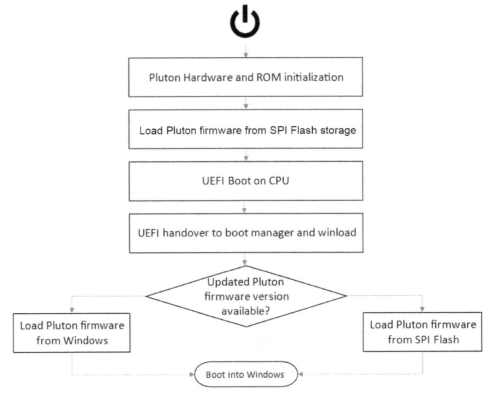

Figure 8.5 – Microsoft Pluton workflow

As shown in *Figure 8.5*, the Pluton hardware initialization is carried out when the system boots. This loads the Pluton firmware from the motherboard's **Serial Peripheral Interface (SPI)** Flash storage. During the Windows startup, the OS uses the copy of the latest firmware version available from Windows (via Windows Update) or uses the loaded SPI Flash version of no newer available.

You can continue your learning of Microsoft Pluton in the following Microsoft Learn article: `https://packt.link/HEfKj`.

In this section, we looked at Microsoft Pluton. We will look at Windows Defender System Guard in the following section.

Windows Defender System Guard

Windows Defender System Guard provides a hardware-based root of trust. It ensures the integrity of a system on boot through device attestation that is both local and remote. It can protect critical resources such as the Windows authentication process, Windows Hello, SSO tokens, and a virtual TPM.

Windows Defender System Guard fits in with the *assumed breach* mindset of our *zero-trust* strategy. We cannot simply trust a system that self-attests to its state of security as it could be potentially compromised. However, we can provide remote analysis of a device's integrity through technologies included in Windows Defender System Guard.

At OS boot, TPM 2.0 is used to take an integrity measurement series through Windows Defender System Guard. Earlier TPM versions such as TPM 1.2 are not supported with System Guard Secure Launch.

To ensure that the measurement data cannot be tampered with in the case of a compromised platform (assume breach), the process and data are hardware-isolated from the OS. The measurements are then used to determine the device's integrity in terms of its firmware, hardware configuration state, the components related to OS boot, and so on.

Once the system has booted, the measurements are sealed by Windows Defender System Guard using TPM devices. A management system such as **Microsoft Endpoint Manager** (which includes **Configuration Manager** and **Intune**) can then acquire these measurements and perform remote analysis. If the integrity of the device is identified as lacking by Windows Defender System Guard, actions can be taken by the management system to deny access to resources by the device.

Windows Defender System Guard has several system requirements, such as a 64-bit computer with a minimum of four cores (logical processors) and TPM 2.0.

The full list of requirements can be found in this Microsoft Learn article: `https://packt.link/0d3YS`.

In the following sections, we will look at System Guard Secure Launch using group policy and the Windows Security app.

Enabling System Guard Secure Launch

This section will briefly examine how System Guard Secure Launch can be enabled and how to verify whether it is running.

Using group policy

In this task, we will enable Secure Launch using group policy:

1. Search for and launch **Edit group policy**:

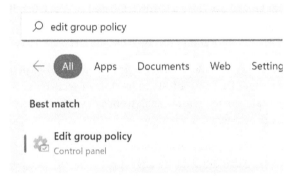

Figure 8.6 – Search and select the option to edit group policy

2. From **Local Group Policy Editor**, click **Computer Configuration | Administrative Templates | System | Device Guard | Turn On Virtualization Based Security**.

3. Then, click on the radio button labeled **Enabled**. Also, under **Secure Launch Configuration**, click **Enabled**, then click **OK**:

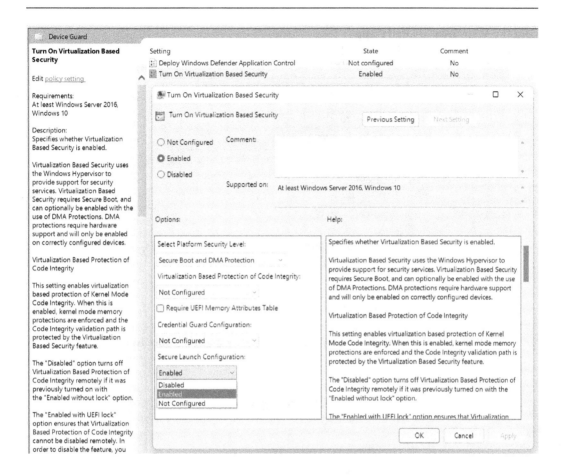

Figure 8.7 – Enable Secure Launch in the local group policy editor

In this task, we enabled Secure Launch using group policy. In the following task, we will use the Windows Security app.

Using the Windows Security app

This task will enable Secure Launch using the Windows Security app:

1. Search for and launch **Windows Security**:

Figure 8.8 – Search and select Windows Security

2. From **Windows Security**, click **Device security | Core isolation | Core isolation details**:

Figure 8.9 – Windows Security app

3. From the **Core isolation** page of the **Windows Security** app, set **Firmware protection** to **On**:

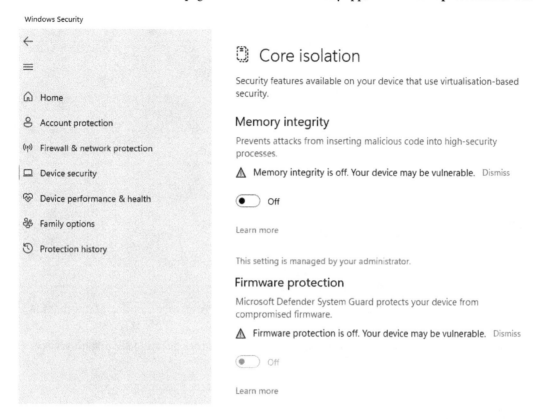

Figure 8.10 – Firmware protection

In this task, we enabled Secure Launch using the Windows Security app. In the following task, we will look at where to find the status of Secure Launch.

Identifying the Secure Launch status

In this task, we will identify the status of Secure Launch:

1. Search for and launch **System Information**:

Figure 8.11 – Search and select system information

2. Locate the following entries and check their status:

   - **Virtualization-based security services configured**: This entry value validates that the services are configured
   - **Virtualization-based security services running**: This entry value validates that the services are running

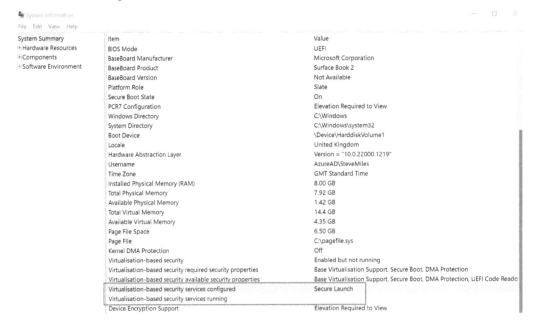

Figure 8.12 – Secure Launch status

In this section, we looked at the Windows Defender System Guard. We will look at Hypervisor-protected Code Integrity in the following section.

Hypervisor-protected Code Integrity

Hypervisor-Protected Code Integrity (**HVCI**) provides security measures at the virtualization layer. It is categorized as **virtualization-based security** (**VBS**) and is a Windows OS feature.

An isolated virtual environment is created using the Windows Hypervisor, which becomes the trust's OS root. It protects against the exploitation of the Windows kernel from malware by using kernel mode code integrity, which runs to restrict kernel memory allocations that could be used for system compromise.

In the following section, we will look at enabling HVCI.

Enabling HVCI

This section will briefly examine how HVCI can be enabled.

Using group policy

In this task, we will enable the HVCI policy:

1. Search for and launch **Edit group policy**.
2. From the **Local Group Policy Editor**, click **Computer Configuration | Administrative Templates | System | Device Guard | Turn On Virtualization Based Security**.
3. Then, click on the radio button labeled **Enabled**. Under **Virtualization Based Protection of Code Integrity**, select **Enabled with UEFI lock** or **Enable without UEFI lock**, then click **OK**:

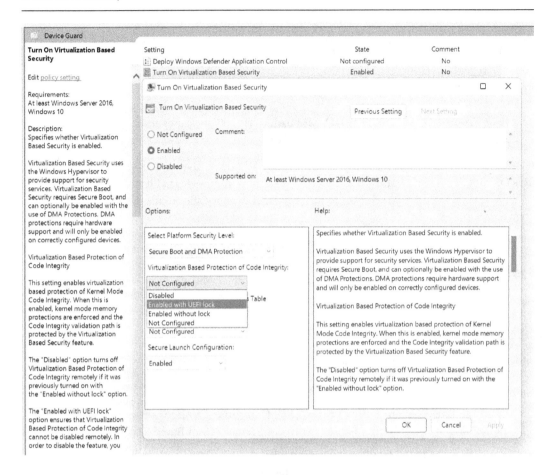

Figure 8.13 – Enabling HVCI in the local group policy editor

Enabling the UEFI lock ensures that the feature cannot be disabled. It is locked.

In this task, we enabled HVCI using group policy. In the following task, we will use the Windows Security app.

Using the Windows Security app

In this task, we will enable HVCI using the Windows Security app:

1. Search for and launch **Windows Security**.

2. From the **Windows Security** app, click **Device security** | **Core isolation** | **Core isolation details**:

Figure 8.14 – Windows Security app

3.   From the **Core isolation** page of the **Windows Security** app, set **Memory integrity** to **On**.

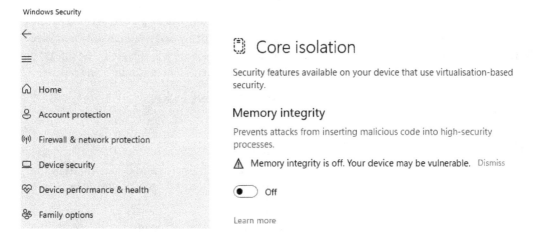

Figure 8.15 – Memory integrity

In this task, we enabled HVCI using the Windows Security app.

Additional configuration references can be found in the following Microsoft Learn article: https://packt.link/qgAlC.

This section on HVCI concluded our look at hardware security. In the following section, we will look at how we can ensure that we operate system security.

Ensuring that we operate system security

As a part of our defense-in-depth look at securing Windows 11 systems, this section looks at the security measures we can take to protect the OS. These include **Secure Boot** and **Trusted Boot**, the Windows Security app, encryption, security baselines, and **Defender**, which we will discuss in the following sections.

Introducing Secure Boot and Trusted Boot

Secure Boot and Trusted Boot work together to provide **OS-level protection** of a Windows device during startup, preventing the loading of malware and corrupted components.

The initial boot-up protection is carried out by Secure Boot. The firmware is verified that it is digitally signed, and then all code that runs before the OS is checked by Secure Boot.

The digital signature of the **OS bootloader** is then checked to ensure the Secure Boot policy will trust it and that there has been no tampering.

Trusted Boot then picks up the process. The digital signal of the Windows kernel is verified by the **Windows bootloader**. Every other Windows startup process component is then verified by the Windows kernel. This includes the boot drivers, startup files, and any **early-launch anti-malware (ELAM)** driver. Any corrupted component detected by the bootloader will not load. The integrity of the device can often be remediated through corrupted components automatically being repaired.

In this section, we looked at Secure Boot and Trusted Boot. We will look at the Windows Security app in the following section.

Exploring the Windows Security app

Security posture is provided by an at-a-glance view of the security status and device status in the Windows Security app. Actionable insights are provided to stay protected by quickly identifying issues and providing actions for remediating them.

You can open the Windows Security app from the notification area of the taskbar, as seen in the following figure:

Figure 8.16 – Taskbar notification area

You can also use the search function to find and launch the **Windows Security** app, as seen in the following figure:

Figure 8.17 – Search for the Windows Security app

The **Windows Security** app, when launched, will be as shown in the following figure:

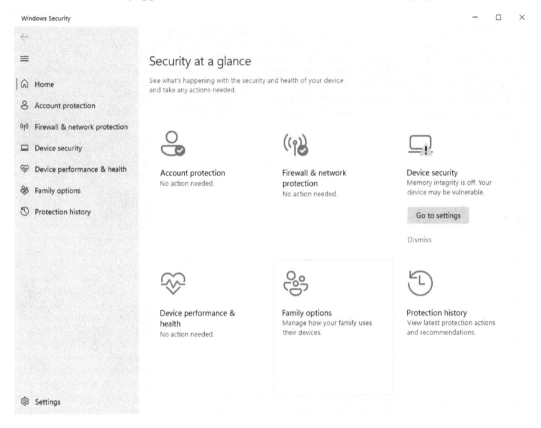

Figure 8.18 – Windows Security app

The following security options are included in the Windows Security app; these sections can also be hidden with a group policy:

Account protection

This section provides information and settings for your account protection and sign-in. This can be seen in the following figure:

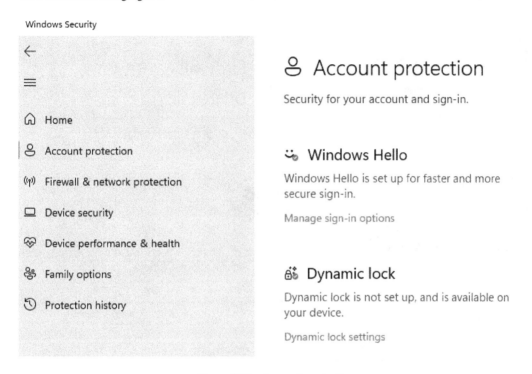

Figure 8.19 – Account protection

Firewall and network protection

This section provides information and settings for firewalls such as Microsoft Defender Firewall or third parties and the device's network connections. This can be seen in the following figure:

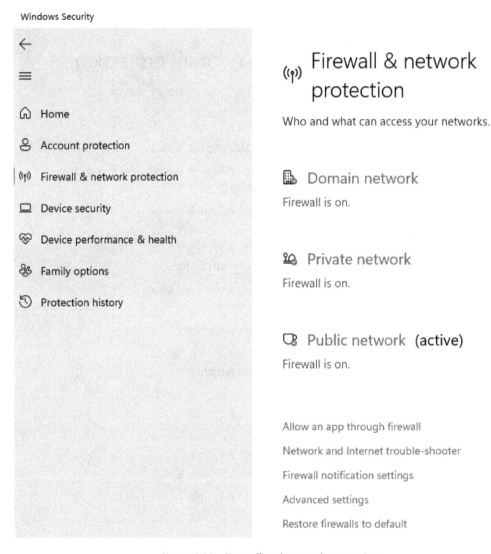

Figure 8.20 – Firewall and network protection

Device security

This section provides information and settings for your device's built-in security. This can be seen in the following figure:

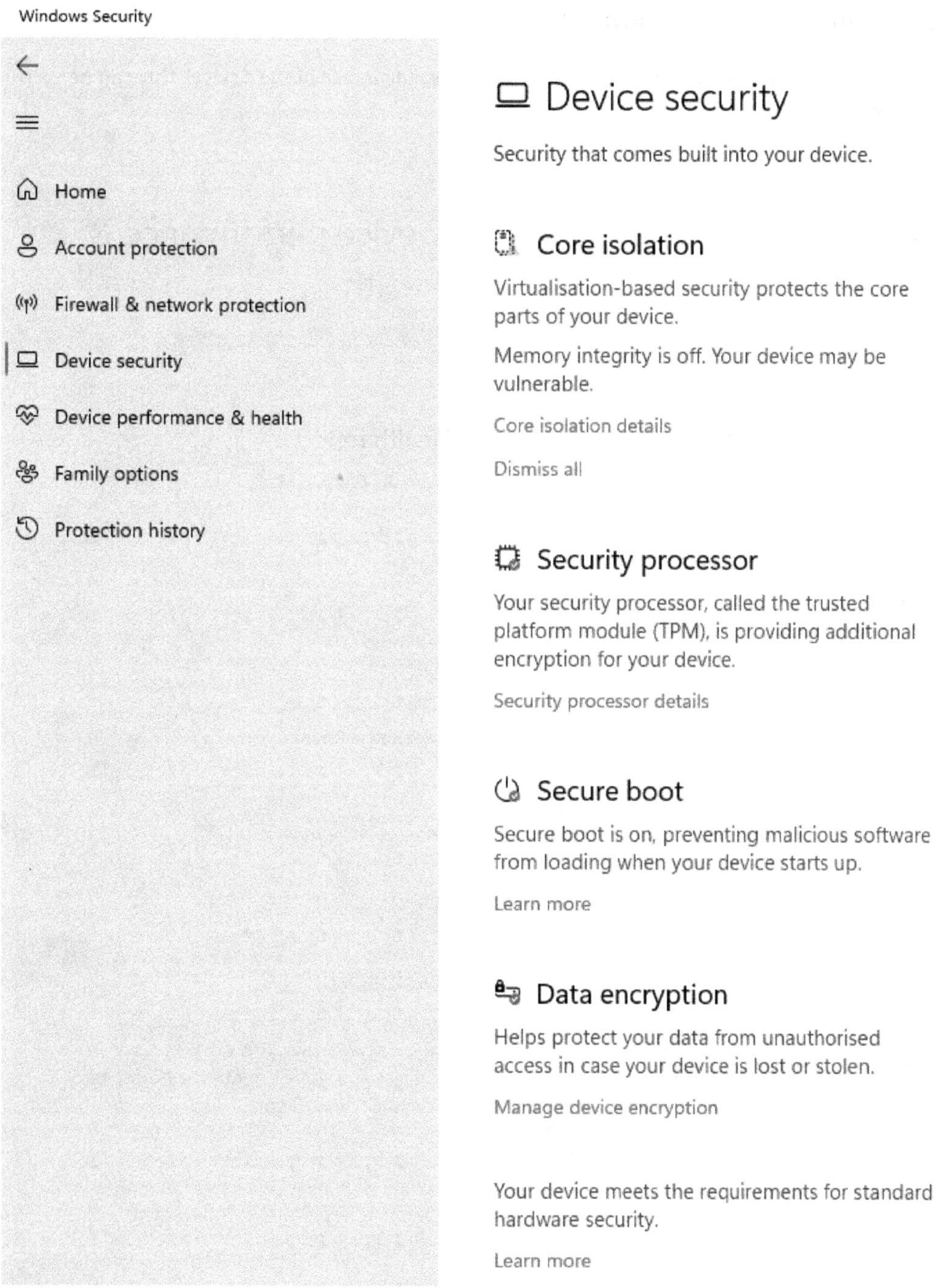

Windows Security

←

≡

⌂ Home

⌂ Account protection

((ᵗ)) Firewall & network protection

▯ Device security

♡ Device performance & health

⚙ Family options

⟲ Protection history

▢ Device security

Security that comes built into your device.

▢ Core isolation

Virtualisation-based security protects the core parts of your device.

Memory integrity is off. Your device may be vulnerable.

Core isolation details

Dismiss all

▢ Security processor

Your security processor, called the trusted platform module (TPM), is providing additional encryption for your device.

Security processor details

⏻ Secure boot

Secure boot is on, preventing malicious software from loading when your device starts up.

Learn more

▣ Data encryption

Helps protect your data from unauthorised access in case your device is lost or stolen.

Manage device encryption

Your device meets the requirements for standard hardware security.

Learn more

Figure 8.21 – Device security

Device performance and health

This section provides information about the health and performance of the device. This can be seen in the following figure:

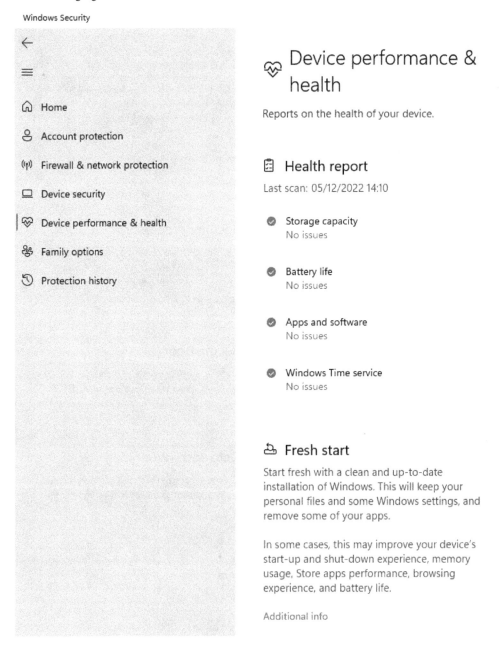

Figure 8.22 – Device performance and health

Family options

This section provides information and settings for parental control; this is not intended for a work environment. This can be seen in the following figure:

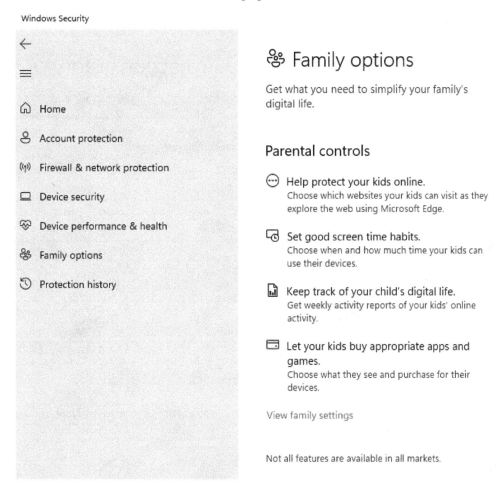

Figure 8.23 – Family options

Protection history

This section provides information for the recommendations and protection actions. This can be seen in the following figure:

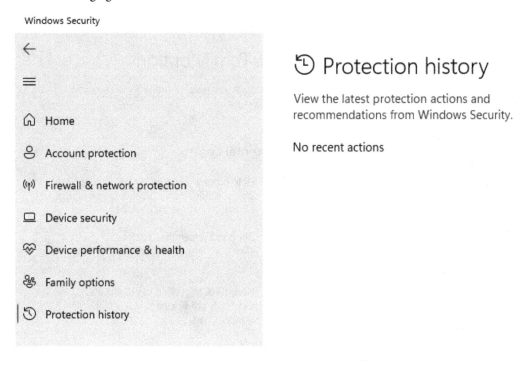

Figure 8.24 – Protection history

In this section, we looked at the Windows Security app. In the following section, we will look at how OSs can be protected by using encryption with BitLocker.

Using BitLocker for encryption

BitLocker provides OS-level integration for data protection through drive encryption. Encryption can be provided for the OS, fixed data, and removable data drives. It provides an extra layer of protection for data access and ensures that data is inaccessible for decommissioned or recycled machines protected with BitLocker.

When combined with **hardware-level security** such as TPM, maximum protection is achieved. There is the option to provide a **multi-factor authentication** mechanism that ensures that a device won't start or resume from hibernation until a device that contains a startup key, such as a USB, is inserted or a user supplies a PIN.

For an OS without TPM, a USB startup key is required to start the device or resume hibernation. However, there is no system integrity verification provided at pre-startup with this option as there would be with TPM.

Further information on many aspects of BitLocker can be found in the following Microsoft Learn FAQ article: `https://packt.link/KpTLp`.

In this section, we looked at encryption using BitLocker. We will look at Microsoft's security baselines in the following section.

Security baselines

Security baselines are a set of Microsoft's recommended security settings – configurations that are considered an industry standard and included as part of Microsoft's **Security Compliance Toolkit (SCT)**. These baselines help you navigate complex security controls and features that a company may require to comply with their industry standards, compliance, and so on.

Due to the size of this topic, it is beyond the scope of this book to dive into more detail so that more information can be found for the SCT at the following Microsoft Learn article: `https://packt.link/LQZNp`.

This section on security baselines concluded our look at OS security. In the following section, we will move on to securing the identity of users.

Ensuring user identity security

In our defense-in-depth look at securing Windows 11 systems, this section looks at the security measures we can take to protect user identity security. This section will cover **Windows Hello for Business** and **Microsoft Defender Credential Guard**.

Windows Hello for Business

Windows Hello for Business is a secure authentication solution that uses **two-factor authentication** on devices to replace passwords. The two factors used for authentication are a **device-tied user credential** and a **biometric** or **PIN**. A PIN is more secure than a password as it is tied to the device.

The following problems with passwords are addressed with Windows Hello:

- It's difficult to remember strong passwords, leading to reuse across sites
- Passwords can be exposed upon breach/phishing attacks
- Replay attacks on passwords

Windows Hello authenticates user identities to allow them access to the following:

- Microsoft Account
- Microsoft Active Directory Account
- Microsoft Azure Active Directory Account
- FIDO-supported identity provider services

Windows Hello provides biometric authentication based on fingerprint, face, and iris recognition.

- **Fingerprint recognition**: This biometric recognition type scans your finger using a capacitive sensor. It can be integrated into either a system or an external device, such as USB keyboards.
- **Facial recognition**: This biometric recognition type uses special IR light cameras. It can differentiate between a person and a scanned picture or photograph. It can be integrated into devices or can be external.
- **Iris recognition**: This biometric recognition type uses a camera to carry out an iris scan. An example is *HoloLens 2*, which includes an iris scanner.

What's the difference between Windows Hello and Windows Hello for Business?

Windows Hello is aimed at *individual/consumer* devices. The login process differs from Windows Hello for Business, as it still uses a password hash.

Windows Hello does not use *certificate-based authentication* or *public/private keys*. However, as a consumer or individual, you are not joined to a domain, so this is a reduced risk.

In contrast, Windows Hello for Business is aimed at the work environment and companies where devices and Windows Hello for Business can be managed by group or MDM policy. As such, PIN authentication is always **key-based** or **certificate-based**.

In this section, we looked at Windows Hello for Business. We will look at Microsoft Defender Credential Guard in the following section.

Microsoft Defender Credential Guard

Microsoft Defender Credential Guard provides *secrets isolation* for access to *privileged system software* only and uses VBS to offer protection for credentials.

Credential Guard prevents credential theft attacks, such **Pass-the-Ticket** or **Pass-the-Hash**, by protecting the hashes of passwords, Kerberos **Ticket-Granting Tickets** (**TGTs**), and domain credentials stored by applications.

> **Note**
>
> Credential Guard is turned on by default, starting in Windows 11 Enterprise, version 22H2, and Windows 11 Education, version 22H2.

How does it work?

Before Credential Guard was introduced, the **Local Security Authority** (**LSA**) would have stored secrets and would be used by the OS in its process memory.

With the introduction of Credential Guard, now the OS LSA process talks to the **Isolated LSA** (**LSAISO**) process when Credential Guard has been enabled. This new process stores and protects the secrets.

VBS protects the LSAISO process data, and the rest of the OS cannot access it. **Remote Procedure Calls** (**RPCs**) are used for LSAISO process communication.

This process does not involve the hosting of any drivers for security reasons; only a small subset of OS binaries are signed with a VBS-trusted certificate. Before launching in a protected environment, the signatures are validated.

The following figure represents the LSA isolation using VBS:

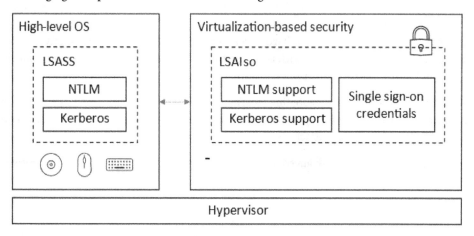

Figure 8.25 – LSA isolation using VBS

Kerberos doesn't allow **unconstrained Kerberos delegation** or **DES encryption** when Credential Guard is enabled; this is the case for signed-in credentials, saved, and prompted credentials.

When Credential Guard is used, single sign-on will not work with NTLMv1, MS-CHAPv2, Digest, and CredSSP protocols. It is not recommended to use valuable credentials with these protocols.

What are its requirements?

The following hardware and software requirements must be met to use Credential Guard:

- Support *required* for virtualization-based security
- Support *required* for Secure Boot
- Support *preferred* for TPM
- Support *preferred* for the UEFI lock

Credential Guard can support physical and Hyper-V virtual machines.

Further information on the requirements can be found in this Microsoft Learn article: `https://packt.link/MnQNT`.

This section on Credential Guard concluded our look at identity security and this chapter on how we can secure Windows 11.

Summary

We should reiterate the relevance of what has been learned in this chapter by understanding that an attack chain of events can be prevented and disrupted through a zero-trust and defense-in-depth approach. We covered various aspects of hardware security and OS security, particularly Windows 11 security. We concluded the chapter with a section on user identity security, where we looked at Windows Hello for Business and Microsoft Defender Credential Guard.

By implementing these measures and adopting security posture management, we can make an attacker consider an easier attack elsewhere that offers the least resistance by putting multiple obstacles in the attacker's way and increasing their attack costs.

In the next chapter, we will cover advanced topics of configuration for use cases in the enterprise.

9
Advanced Configurations

Windows 11 is an advancement of Microsoft's general direction for Windows clients. It offers improved performance and reliability, along with ease of deployment and enhanced security capabilities, compared to previous versions.

Previous versions of Windows allowed something of a free-for-all mentality in customizing images and Windows installations. It is worth noting that some of the techniques developed by IT professionals outside of Microsoft were not truly supported by Microsoft. These configuration tweaks, however, certainly achieved the goals of the IT professionals to customize the Windows installation for the required business use case. Usually, the solutions were stable (enough), and Microsoft provided best-effort support when issues arose, so things were good.

However, as IT organizations in large enterprises matured, business folks became involved more in the IT process. ISO, **Information Technology Infrastructure Library** (**ITIL**), change review boards, procedures, and so on all came into the IT realm. At this point, the best-effort and stable (enough) aspects of solutions became issues to address. You do not simply run global, enterprise-grade applications that the world depends upon with kludged solutions, hacked together from tips from various blogs and forums found around the internet.

To help IT organizations reduce support incidents, increase the stability of the solutions, and benefit the Windows platform, changes and recommendations started being made. In Windows XP, for example, it was common practice to swap **Hardware Abstraction Layers** (**HALs**) based on CPU architecture so that an organization could use a single image to deploy Windows. Microsoft never really supported this, but looking at the reasoning, the common issue was that Windows didn't handle this properly, so we hacked a solution to get around the issue. Therefore, making Windows not have the issue in the first place became a focus for engineering resources.

So came Windows 8.1, Windows 10, and Windows 11, which presented a conscious lock-down of the standard user areas of the operating system and isolation of the user experience configuration files. The idea is to train users to not store things in `C:\somepath` but, instead, in the user profile, for example. On the technical tools, the goal was to train IT staff to image Windows in a reliable, repeatable method by creating the **Microsoft Deployment Toolkit** (**MDT**) and **System Center Configuration Manager** (**SCCM**).

In this chapter, we will cover the following topics:

- Virtual desktops

- The Windows Configuration Designer

- Windows 11 Kiosk Mode

- Windows Autopilot

- The Set up School PCs application

- Unbranded Boot

- Unified Write Filter

- Windows Subsystem for Linux

- Group Policy Editor

- Remote Desktop Protocol

- Windows Hello and Windows Hello for Business

- Window Firewall with Advanced Security

- Hyper-V

- Windows Task Scheduler

- Enabling BitLocker drive encryption

- Storage Spaces Direct

- Windows Defender Application Guard

Virtual desktops

Windows 11 represents a convergence of methods to access applications and enterprise data. While **Virtual Drive Infrastructure** (**VDI**) has been ubiquitous in most enterprise environments for some time, the method of configuring the Windows image has been an iterative improvement process. Azure Virtual Desktop, Microsoft 365, and Microsoft Dev Box are offerings from Microsoft that represent different approaches to the virtual desktop space:

- Azure Virtual Desktop is an offering from Microsoft that can be run in Azure or on-prem

- Microsoft 365 is a pre-configured, cloud-based service that provides a desktop to users from anywhere

- Microsoft Dev Box is a managed service that allows developers to spin up ready-to-use virtual machines that are pre-configured for a specific product or project

These offerings bring VDI access to organizations that don't have the expertise or up-front resources needed to run on-prem VDI farms, allowing for greater flexibility in how a user uses Windows.

On-prem virtual desktop best practices

On-prem VDI projects require considerable underlying and complementary infrastructures to work well. The scale and scope of these projects are themselves left to other books. Further, it is important to also be able to speak about these subjects and interact with the team or teams managing the infrastructure that a VDI project runs on top of, as troubleshooting and configuration/architecture of these items will play a part in the success or failure of the project.

The backend storage assigned to a project should be excellent. Ideally, it would be architected with the golden image on flash storage, with differencing disks linked back to the golden image. The differencing disks would also reside in a flash memory pool.

High-speed 10+ GB Ethernet or fiber channel connectivity would be necessary as well. It has been my experience that CPU contention is generally the bottleneck in performance. One consideration that often goes unthought of is system board and CPU location of the hypervisor host, where a virtual machine's CPUs are resident on multiple physical processor sockets on the host system. The communication from one core to another in the VM is then slowed, as it must pass along the **QuickPath Interconnect** (**QPI**) bus (for more information, see https://packt.link/Myefw). Quality hypervisors will enable you to disallow NUMA spanning for VMs, or allow the admin to configure systems for specific workloads.

Another consideration is whether user scenarios will require GPU acceleration to view videos at a high quality (stock traders watching the news, for instance), allow engineers to run AutoCAD, allow data scientists to work with GPU-enhanced workloads, and so on.

If so, the traditional blade server chassis used to achieve maximum density may not suffice for your host configuration. This expands the cost of the system, as density per host usually decreases in this scenario, and the chassis with GPUs tend to be larger, generate more heat, and consume more power, all of which add to the final operating cost of the solution.

The **User Profile Disks** (**UPDs**) are another factor to consider, as well as application virtualization or layering technologies. Simply put, UPDs are a way of "bubbling" the user profile (e.g., settings, preferences, links, and data) into an object that the VM streams and presents as part of the VDI experience.

Application virtualization is a technique whereby an application is bubbled or otherwise contained as a unique entity unto itself. It still requires the OS to run, and so forth, Then, the application can be streamed into the image and launched at request time, versus being baked into the master golden image. There are pro and con arguments for this that could take up a whole chapter or two. Sometimes this technology fits, while sometimes it appears silly and cumbersome.

Layering technologies from Citrix and VMWare allow a similar approach to application virtualization. However, instead of each app sort of streaming in, when the user logs on to the guest image, the app is presented based on the profile of the user. *What apps does the user need? Oh A, B, and C? Okay, let's provide the base OS layer, then layer on top of the strata additional application layers.* This sounds odd at first, but the technique is quite fast and efficient. User profile disks are a way to bubble the user data, documents, downloads, desktop folders, settings, registry data, and so on of the user into a unified experience, one might think of it as a container for the user data. This is then provided at logon to the user.

VDI configurations

With all this information at hand, how does a system administrator create an image for Windows 11 that will run in this VDI configuration? Generally, in previous Windows versions, administrators used vendor-provided scripts or tools provided by Citrix or VMWare to configure the Windows image to make it ready to become a virtual machine. With Windows 11, that certainly can still be the case. Microsoft has desktops for Windows 11 in Azure, and AWS has a similar offering as well. With these types of configurations becoming more common, guidance for VDI tuning is also documented at `https://packt.link/n0on7`.

Perhaps one of the often-overlooked considerations for a VDI deployment is the ability to collect diagnostic data from virtual guests. How will you collect a full memory dump if required? While AVD Collect exists for that specific offering from Microsoft, other use cases of VDI tend to increase the support burden, by requiring support calls to involve multiple vendors all working together to help the customer succeed. An example of this would be the performance impact of applications and monitoring software, such as data loss prevention suites. These are things that ideally are thought through prior to an emergency call with a Microsoft or other support engineer. In conclusion, the key point for a VDI solution is to spend money to do it right.

The Windows Configuration Designer

In the past, it was common for an IT administrator to purchase a computer with a perfectly good Windows OS preinstalled and then bring it into the corporate environment by re-imaging it with a corporate image. The reasons for this vary from organization to organization but often entail the usual suspects of **Original Equipment Manufacturer** (**OEM**)-installed cruft, trial applications, or even the wrong SKU of Windows. This is a rather inefficient process that needlessly throws out the whole operating system when only a small subset needs to be reconfigured.

The Windows Configuration Designer is one method to create a repeatable process for image customization. With it, you can create provisioning packages to easily configure Windows 11. It can be used by IT admins to provision either BYOD or business/educational assets.

Windows 11 Kiosk Mode

Windows 11 Kiosk Mode is a feature of Windows 11 that is designed for use in limited security or multi-user environments, restricting access to a single application or set of applications. In a scenario such as an interactive directory in a building lobby, a device will need to provide the building directory functionality to many users without requiring the users to authenticate. It will also need to restrict users from accessing any applications outside of the directory application.

In order to accomplish this, Kiosk Mode replaces Windows Explorer, the default shell, with an alternative shell and limited application access, specified by the administrator. When the replacement shell application is closed, the user session is ended, so there is no way to access the underlying operating system.

Provisioning Kiosk Mode can be accomplished in a number of ways, depending on the type of application that is chosen as the Kiosk app and the edition of Windows 11 that is configured. Kiosk Mode for UWP applications is available in Windows 11 Pro, Education, or Enterprise, while classic Win32 apps are available only for Windows 11 Enterprise or Education.

If a **Universal Windows Platform** (**UWP**) application is used as the kiosk app, the configuration can be accomplished manually using the **Settings** application, or in a more automated fashion, using a PowerShell script. Windows 11 Enterprise or Education also offers you the ability to configure a UWP kiosk app using a **Mobile Device Management** (**MDM**) policy or provisioning package.

To configure a classic Windows app as a Kiosk app, an administrator must use the Windows 11 feature Shell Launcher, which will need to be installed using the programs and features wizard or **Deployment Image Servicing and Management** (**DISM**). Once enabled, the `root\standardcimv2\embedded` WMI namespace can be used to configure the default shell for a user or group, as well as the action that should be performed when the shell application exits. These actions include relaunching the shell application, restarting the computer, and shutting the computer down.

Windows Autopilot

Windows Autopilot is, in a sense, a system management and deployment tool, but without servers. Similar to Microsoft's Intune or SCCM, Windows AutoPilot can be used to manage devices. It requires Azure AD and some cloud-based services, but the result is you can configure and tweak your devices and recover/reconfigure them quite easily, without the infrastructure costs associated with a traditional SCCM multi-site deployment architecture. Autopilot's features are as follows:

- Automatically joining devices to Azure AD
- Auto-enrolling devices into MDM services, such as Microsoft Intune, which requires an Azure AD Premium subscription: `https://packt.link/dKfuN`
- Restricting administrator account creation

- Creating and auto-assigning devices to configuration groups based on a device's profile
- Customizing **Out of Box Experience (OOBE)** content specific to an organization

The reason to use Autopilot is to pre-configure Windows PCs (or HoloLens 2 devices) and speed deployment times, without any investment in on-prem infrastructure, thereby reducing the IT cost to deploy endpoints to end users.

The Set up School PCs application

Microsoft has an application somewhat akin to the Windows Configuration Designer, named Set up School PCs. This is an application for teachers or school IT staff to set up machines for student use. While the application was last updated in 2021, it does still work with Windows 10 and 11/11 SE.

The application allows an administrator to configure specific lockdown policies for Windows for Education systems, such as Custom Logon, Shell Launcher, and Keyboard Filter (discussed in the *Device lockdown* section). It also can automate joining the computer to Intune for Education and configure the system to take tests and the like.

You can install the tool from the Microsoft Store and then use it to configure USB sticks, deploying configuration settings via USB.

Figure 9.1 – Microsoft Store – the Set up School PCs application

It's OK at what it does; think of it as somewhat akin to the Windows Media Creation Tool but with additional configuration options.

Device lockdown

Windows 8 was the last Microsoft OS to deliver an embedded edition as a formal SKU. In Windows 11, the customization can be applied directly to a Windows 11 Enterprise, Education, or IoT installation (or image file). The customization available to modify an image is found in the Windows features, under the device lockdown category. Those who have crafted images for Windows Embedded in the past will recognize the options and be familiar with the capabilities already. Certainly, they are worth covering in this chapter for a clear understanding of what these capabilities are and aren't.

These features are available only in Windows 11 Enterprise and Education editions, yet they are visible on a Windows 11 Professional installation. However, they may or may not work properly on Professional and likely violate license terms if they do. Device lockdown consists of the following checkbox options:

- **Custom Logon**: Defined as *"enables customized logon experiences,"* which is a bit vague
- **Keyboard Filter**: Prevents unwanted keystrokes (enabling audit mode, avoiding auto-login, and so on)
- **Shell Launcher**: For launching a Windows application as a shell
- **Unbranded Boot**: Suppresses Windows elements that appear when Windows starts, resumes, or encounters errors
- **Unified Write Filter**: Installs services and tools to protect physical media from write operations

Note that checking these boxes enables a feature for configuration; additional work is needed to implement the feature properly. An overview of each option is found next:

- **Custom Logon**: Custom Logon can modify the logon screen to remove certain system components.
- **AnimationDisabled**: This disables the animation screen displayed during a new Windows logon.
- **Shell Launcher** Shell Launcher allows an administrator to specify an application as the shell in Windows. When the application is closed, the system logs out.
- **BrandingNeutral**: This is a setting that can disable the power button, language button, ease of access, switch user, and so on.
- **HideAutoLogonUI**: If you enable automatic sign-in for a device, this will hide or show the Windows welcome screen during the logon process.
- **NoLockScreen**: Useful for kiosks and other devices, this setting will be set if the lock screen is invoked when a machine is idle.
- **Keyboard Filter**: This one is straightforward. This filter is used to prevent keys such as *PrtScrn* and combinations such as *Ctrl + Alt + Delete* from doing their normal functions. For managed workstations or stations that may be in unknown people's hands (think kiosks at airports, ATMs, medical devices, and so on), this is a great way to lock them down to prevent tampering.

These settings give the enterprise administrator additional flexibility to tweak their image and prevent tampering, as well as securing the endpoint and applying branding to it.

Unbranded Boot

Unbranded Boot lets a system administrator customize items such as the boot loading screen and animations, as well as shutdown experiences. The primary use case is, naturally, device branding.

The specific registry keys are as follows:

- **DisableBootMenu**: To prevent tampering, the administrator can disable *F8* and *F10* keys during startup

- **DisplayDisabled**: If your system encounters an error, it displays a blank screen instead of the Microsoft image

- **HideAllBootUI**: This hides all the Windows UI elements during boot (the logo, scrolling status indicator, status messages, and so on)

- **HideBootLogo**: This suppresses the Windows boot logo at boot

- **HideBootStatusIndicator**: This setting suppresses the status indicator displayed during boot

- **HideBootStatusMessage**: This, similar to **HideAllBootUI**, hides the boot status of the OS loading phase (applying Group Policy messages and so on)

- **CrashDumpEnabled**: This setting has a few values that govern the size of a dump captured when a system encounters a stop condition.

You would configure these settings to brand a device with a corporate logo or to control other aspects of a system.

Unified Write Filter

A **Unified Write Filter** (**UWF**) is a filter driver that seals a drive in a non-write view and then keeps a differencing area in RAM of all the changes a user makes during the session. This area is known as the UWF overlay. This is a virtual storage area that looks at all the intended writes for the protected storage area. Instead of performing the write, it reads that disk sector from the disk and then modifies it as the write was supposed to, holding that change and caching it in memory (unless a paging file is in use; then, it can make use of the page-file to extend the overlay area).

This is the biggest drawback of UWF. Typically, embedded devices do not have a preponderance of RAM installed (they are supposed to be cheaper than desktops, after all) and their storage is slow as well, so if a user does too much on the device, you run the risk of actually running out of RAM on the device.

To mitigate this, you can exclude areas of the storage from UWF protection (much like an antivirus). Administrators must do this, and a restart of the device is required for them to take effect.

The same consideration is made for the Windows registry as the disk volume, and areas can be excluded there as well.

Applying updates to a UWF-protected volume some additional steps that are well documented at `https://packt.link/IuOck`.

Windows Subsystem for Linux

Windows 11 has the feature **Windows Subsystem for Linux (WSL)**, which allows an end user to run Linux command-line tools and utilities directly on Windows. Windows Subsystem for Linux comes in two versions, WSL and WSL2.

The key difference between the two is that WSL2 uses Hyper-V architecture, provides access to the GPU, and improves the performance of the Linux instance. At the time of writing, there is no plan to deprecate WSL, and you can even run WSL and WSL2 side by side. To learn more, visit `https://packt.link/IjNoz`.

To enable WSL, do the following:

1. Open **Start** and search for `Windows Features`; from there, select **Turn Windows features on or off**.

2. Once in the **Windows Features** window, scroll to find **Windows Subsystem for Linux**, and check the box next to it.

3. To start the installation process, click on **OK**; this may take a few minutes to complete and may prompt you to reboot your computer.

4. After the installation is complete, visit the Microsoft Store to search for a Linux **Distribution (Distro)**, such as Ubuntu or Debian.

5. Click on the **Get** or **Download** button to install the distribution.

6. Once installed, search in the **Start** menu for the Linux distribution, and click on the icon to start it.

7. Complete the setup process, which may include creating a new user account and password.

Then, you can run Linux commands and utilities directly on Windows as well as retain access to Windows filesystems from within the Linux environment.

Group Policy Editor

The **Group Policy Editor (GPE)** is a tool that allows you to control settings and configure policies for users and computers in a Windows domain. This is done to provide security or operational configuration changes to endpoints during a domain join, or afterward to endpoints managed by AD/AAD. These settings can be filtered to criteria such as the chassis type, what site the endpoint is at, or even what user currently uses the device.

The GPE is a tool that allows the control of settings and policy configurations for computers and users in a Windows domain. Here are the steps to utilize the GPE:

1. Open the **Start** menu to search for **Group Policy Editor**, and select **Edit Group Policy** from the results list.

2. Once in the GPE, navigate to the **Organizational Unit (OU)**, domain, or local computer that is to be configured. The left pane will display different categories of policies, such as **User Configuration** and **Computer Configuration**.

3. Select a category and subcategory of the policy you want to configure. For example, the **Computer Configuration** category contains **Windows Settings**, which in turn contains **Security Settings**. On the right pane, you can find the available policies and settings available for configuration.

4. To enable or disable a policy, simply double-click on it.

5. Once you have configured the policies, click on the **File** menu and then the **Save** option.

6. To link a GPO to a specific OU, domain, or site, right-click your choice, select **Link to an Existing GPO**, and select the GPO you want to link.

7. To check what settings are applied to a specific computer or individual user, use the **Resultant Set of Policy (RSOP)**.

8. If you want to remove the GPO from a specific location, select **Delete** from the menu after right-clicking the GPO.

An important note is that GPO management requires a certain level of access and permissions. It is recommended to have a good understanding of the GPO settings and to test the policies in a test environment before applying them to production. Also, be aware of the dependencies of the policies you modify, as modifying certain policies can have unintended consequences on the system or other policies.

Remote Desktop Protocol

Windows' built-in Remote Desktop feature allows the remote access and control of another Windows machine over a network or the internet. To configure Remote Desktop on Windows 11 and allow remote access, follow these steps:

1. Search for and then select **System Properties** in the **Start** menu.

2. In the **System Properties** window, select **Remote**.

3. Once in Remote Desktop, click the **Allow remote connections to this computer** checkbox. If you want only specific users or groups to be allowed connections, click on the **Select Users** button to add them. If you want to allow connections from any user, click the **Allow connections from computers running any version of Remote Desktop** checkbox.

4. To save the changes, click **Apply**.

5. On the computer you want to use to connect remotely, search for `Remote Desktop Connection`, and then select **Remote Desktop Connection** from the **Start** menu.

6. In the window for **Remote Desktop Connection**, input the hostname or IP address of the computer you want to connect to in the **Computer** field.

7. Click on **Connect** to establish a remote connection. If prompted with a further login request, enter the login credentials of an account from the computer with remote access permissions.

It's important to note that **Remote Desktop Protocol (RDP)** is not secure by default; therefore, it is recommended that Remote Desktop is used with a **Virtual Private Network (VPN)** or **Remote Desktop Gateway (RDG)** to encrypt traffic and authenticate a user prior to the RDP session, with **Network Level Authentication (NLA)**. Also, it's important to keep Windows up to date, as RDP vulnerabilities have been exploited before.

Windows Hello and Windows Hello for Business

Windows Hello is a feature in Windows 11 that uses biometric authentication instead of using a password. This allows a user to scan their face, fingerprint, or iris to log in to their computer. To set up Windows Hello on a computer, follow these steps:

1. Search for **Settings** in the **Start** menu, and select **Settings** from the results listed.

2. Select **Accounts** from the **Settings** window.

3. Under **Sign-in options**, select **Windows Hello**.

4. Choose the biometric option you want to set up, such as fingerprint or face.

5. Follow the prompts to set up the biometric option, which may include scanning your fingerprint or taking a picture of your face.

Windows Hello and Hello for Business both improve the security of the endpoint. The key difference between the two is that Windows Hello for Business supports certificate-based authentication and single sign-on, helping to reduce the number of password resets and so on.

Studies have shown that biometric logons with encrypted data reduce administrator overhead associated with user account maintenance tasks. Windows Hello and Hello for Business are natural ways to utilize this feature.

Windows Firewall with Advanced Security

Windows Firewall is a security feature built into Windows that works to protect a Windows computer from unauthorized network access. Here are the steps to configure Windows Firewall:

1. Search in the **Start** menu for `Windows Firewall with Advanced Security`, and select it from the results list.

2. Click on **Inbound Rules** or **Outbound Rules** on the left pane of the **Windows Firewall with Advanced Security** window to select which rule you want to create.

3. Click on **New Rule** in the right pane. Select the type of rule you want to create in the **New Inbound** or **Outbound Rule** wizard, such as **Port**, **Program**, or **Predefined**.

4. After the rule is created, you can enable or disable it by selecting it and clicking on **Enable Rule** or **Disable Rule** on the right pane.

A realistic use case for this configuration would be for a company that wants to block all incoming traffic from a specific IP address or IP range. For example, an organization can create a new inbound rule to block all traffic from a specific IP address that is known to be associated with malicious activity. The organization can set the rule to block the traffic on the specific IP address, and it can also set a notification for the IT administrator if there is an attempt to connect from that IP.

It is important to note that configuring Windows Firewall can have unintended consequences if not executed properly, so it is recommended to have a good understanding of the network and the traffic patterns before making any changes.

Hyper-V

Hyper-V is Windows' built-in virtualization feature that allows a user to create and run virtual machines on their computer. Follow these steps to install and set up Hyper-V on a Windows computer:

1. Search for `Windows Features` in the **Start** menu, and click **Turn Windows Features on or off** from the results listed.

2. Scroll down in the **Windows Features** panel to find and check the box next to **Hyper-V**, and then click on **OK** to start the installation process. This may take a few minutes to complete and may present a reboot computer prompt.

3. Once the installation has been completed, search for and select **Hyper-V Manager** in the **Start** menu.

4. In the **Hyper-V Manager** window, a new virtual machine can be created by clicking **New** and then choosing **Virtual Machine** from the menu. The virtual machine can then be named and configured by following the prompts, including specifying the amount of memory and virtual hard drive size of the virtual machine.

    You can also use the quick start feature in Hyper-V to quickly provision an environment.

5. Once you have created the virtual machine, right-click on it, and select **Start** from the menu to start it. A bootable ISO file or physical media will be required to install an operating system on the virtual machine.

6. Use the **Connect** button to connect to the virtual DVD drive on the virtual machine, in order to mount the ISO file or physical media.

7. Once the operating system is installed, you can use Remote Desktop or the Hyper-V console to connect to the virtual machine and configure it as you would a physical one.

It's important to note that Hyper-V requires a certain level of hardware virtualization support, and it may also have an impact on the performance of the host machine, depending on the resources that are allocated to the virtual machines. Also, it is important to keep both the Hyper-V host and the virtual machine guests updated to avoid vulnerabilities and maintain optimal performance.

Windows Task Scheduler

The Task Scheduler is a feature in Windows that allows a user to create, manage, and automate scripts and tasks on their computer. To use the Task Scheduler on a Windows computer to create a task, follow these steps:

1. Search for `Task Scheduler` in the **Start** menu, and choose **Task Scheduler** from the list of results.

2. Create a new task in the **Task Scheduler** window, by either clicking on the **Create Basic Task** button or right-clicking on **Task Scheduler Library** and selecting **New Basic Task**.

3. Follow the prompts to configure the task, including specifying the name, trigger (when the task should run), and action (what the task should do).

4. Once the task is created, you can enable or disable it by selecting it and clicking on the **Enabled** or **Disabled** button.

A realistic use case for this configuration would be for a company that wants to automatically back up its databases every day at a specific time. The system administrator can create a new task that runs a script that backs up the databases and saves the backups to a specific location. Then, the task can be set to run every day at 11 p.m. along with a notification if there is an error.

It's important to note that tasks created with the Task Scheduler can have unintended consequences if not executed properly, so it is recommended to have a good understanding of the script or task and to test it before scheduling it. Also, it's important to regularly review tasks and remove any unnecessary or outdated ones, to avoid overloading the system.

Enabling BitLocker drive encryption

BitLocker is a built-in encryption feature in Windows that allows you to encrypt the entire drive to protect your data from unauthorized access. Enable BitLocker on a computer running Windows by following these steps:

1. Search for `BitLocker Drive Encryption` in the **Start** menu, and select **Manage Bitlocker** from the results listed.

2. Select the drive you want to encrypt from the list of available drives in the **BitLocker Drive Encryption** window.

3. Click on **Turn on BitLocker** to start the encryption process.

4. Choose how you want to unlock your drive on the next boot, such as with a password, smart card, or device's **Trusted Platform Module (TPM)**.

5. Select the encryption method, such as encrypting the entire drive or just the used space.

6. Select the encryption algorithm you want to use and the encryption key size.

7. Choose whether you want to create a recovery key and where to save it.

8. Begin the encryption process by clicking on **Start Encrypting**; the amount of time this takes will depend on the size of the drive as well as the amount of data on it.

In an enterprise environment, this is stored in either OneDrive for Business, AD, or some other central repository.

Storage Spaces Direct

Follow these steps to enable and set up Storage Spaces Direct on a Windows computer:

1. Search for `Storage Spaces Direct` in the **Start** menu, and select **Storage Spaces Direct Manager**.

2. In the **Storage Spaces Direct Manager** window, click on **Enable Storage Spaces Direct**. This begins the process of enabling Storage Spaces Direct on your computer.

3. Once the process is complete, the drives will be grouped into the storage pool, allowing you to create virtual disks (storage spaces).

4. Click on **New Storage Space** in the **Storage Spaces Direct Manager** window to create a new virtual disk, following the prompts for configuration. This may include specifying the name, size, and storage layout.

5. Once the virtual disk has been created, you can format it and assign it a drive letter; then, you can use it like any other drive.

Note that Storage Spaces Direct requires specific hardware, and it is recommended to have a good understanding of the storage infrastructure and the data that will be stored on it, in order to design the right storage solution. Also, it is important to keep Windows up to date and drives healthy to avoid data loss or unavailability.

Windows Defender Application Guard

Windows Defender Application Guard (**WDAG**) is a security feature in Windows 11 that isolates web browsing sessions in a containerized environment, protecting the host system from malicious websites or scripts.

When a user navigates to a website, WDAG checks the website against a list of trusted sites, and if the website is not on the list, the website will open in a containerized environment, isolated from the host system. This prevents any malicious code or scripts from executing on the host system and accessing sensitive data.

A legitimate use case for this feature would be for an organization that wants to protect its employees from phishing and malicious websites while they browse the internet. For example, a financial institution can enable WDAG on their employees' computers and configure the trusted sites list to include only financial websites; this would prevent employees from accidentally visiting a phishing website that mimics a financial institution's website. This would also provide an additional layer of security to the organization, as it would prevent any malicious code from executing on the host system and accessing sensitive data.

It's important to note that WDAG can have an impact on the performance of the host system, depending on the resources that are allocated to the containerized environment. Also, it is important to configure the trusted site list carefully to avoid blocking legitimate websites.

Summary

In this chapter, we touched upon many different topics relating to customizing and configuring the Windows image for enterprise use cases – point of sale, medical devices, kiosks in public areas, and virtualized desktops, to name a few. Windows 11 can be customized in a variety of ways to meet the needs of a changing world.

In the following chapter, we will discuss the changes from Windows 10 to Windows 11 up to 22H2.

10

Windows 11 21H2 and 22H2 Changes (versus Windows 10)

In this chapter, we want to give an overview of the numerous changes in the first two **Windows 11** versions compared to the previous **Windows 10** versions. Unfortunately, Windows 11 is often only reduced to the modified **Start** menu. Also, the OS itself also *disguises* itself as Windows 10 in many places. However, this is for compatibility reasons, as already explained in *Chapter 1, Windows 11 – Installation and Upgrading*. We do not want to limit ourselves to the obvious changes in the UI. Since some features are still being backported from Windows 11 to Windows 10 or are still being added through *Moments* in Windows 11, this is sometimes a bit of a moving target. This chapter is a good start to familiarize yourself with the new features that enhance the user experience and allow you to work more efficiently with Windows 11. It also gives you an overview of the new security features you should definitely take a look at.

In this chapter, you will learn more about the following topics:

- Major changes in the UI
- New programs
- Changes *under the hood*
- New security options
- New inclusive features
- **Continuous Innovation** (CI) aka Moments
- Deprecated and removed features

Some of the changes also appear twice because they were partly introduced with Windows 11 version 21H2, but also already extended, improved, or changed again with Windows 11 version 22H2. We cannot discuss all the new features in detail, and there are still some added by the Moments extensions, but it is a good start to familiarize yourself with them.

Windows 11 21H2 (build 22000)

The first Windows 11 21H2 version was released in **General Availability (GA)** in October 2021 with build number 22000.194 and initially marketed as **Windows 11** without any addition. Shortly before Windows 11 22H2 was released, it was *renamed* Windows 11 21H2. However, you can still find documents, articles, and references that refer to Windows 11. Therefore, it is best to pay attention to the build number. In the following sections, let's see which features were introduced with Windows 11.

New Start menu and taskbar

The most discussed (and loved or hated) new feature is the completely reprogrammed **Start** menu and taskbar. Due to this reprogramming, some features are no longer available.

The new **Start** menu has a completely different look than the previous Windows 10 **Start** menu. At the first click, you now get an overview of 6x3 app icons (scrollable). This number is only expandable or reduced with Windows 11 22H2. Below that are the last used documents/files (3x2). All other apps are only available after clicking on **All apps** in the upper-right corner. Then, you get an overview sorted by letters similar to the old **Start** menu. **Live Tiles** are no longer available.

The new **Start** menu is not resizable as a whole, neither in 21H2 nor so far in 22H2, which is a pity, especially on large monitors. This is what it looks like:

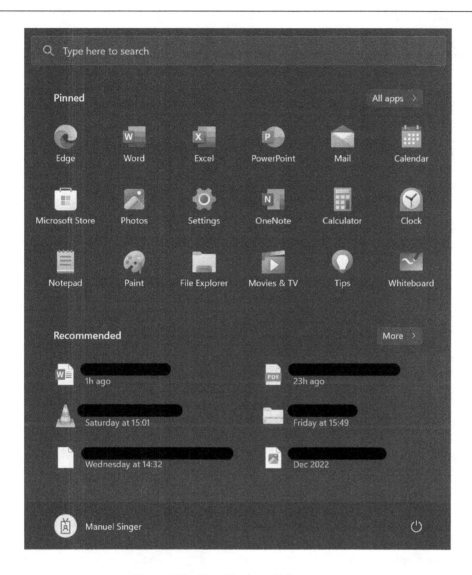

Figure 10.1 – New Windows 11 Start menu

Also, the taskbar has been renewed, but the visual changes are less dramatic. Most prominent is the new **Widgets** section (which was introduced in Windows 10 in summer 2021). In Windows 11, the widgets were first prominent next to the **Start** button; after a **Cumulative Update** (**CU**), they were moved to the outer-left area (if you leave the taskbar at default centered view). The new taskbar can only be docked at the bottom—no more top/left/right alignment. By default, the **Start** button is now centered but you can switch it back to left alignment. But then, the widgets are again next to your **Start** button.

The new taskbar is also missing features such as right-clicking anywhere on the taskbar to access **Task Manager** (use *Win* + *X* as a keyboard shortcut). Also, some grouping options are missing in comparison to the Windows 10 version of the taskbar.

As a new feature, you can now select which contents are displayed on your taskbar when using multiple monitors. The default option is to show all open apps on **All taskbars**, but you can select to only show active windows on that monitor (**Taskbar where window is open**) or a mixture of the main taskbar and open windows (**Main taskbar and taskbar where window is open**):

Taskbar behaviours
Taskbar alignment, badging, automatically hide, and multiple displays   ∧

Taskbar alignment | Centre   ∨ |

☐ Automatically hide the taskbar

☑ Show badges on taskbar apps

☑ Show flashing on taskbar apps

☑ Show my taskbar on all displays

☑ Show recent searches when I hover over the search icon

When using multiple displays, show my taskbar apps on | All taskbars

☑ Share any window from my taskbar   Main taskbar and taskbar where window is open

☑ Select the far corner of the taskbar to show the desktop   Taskbar where window is open

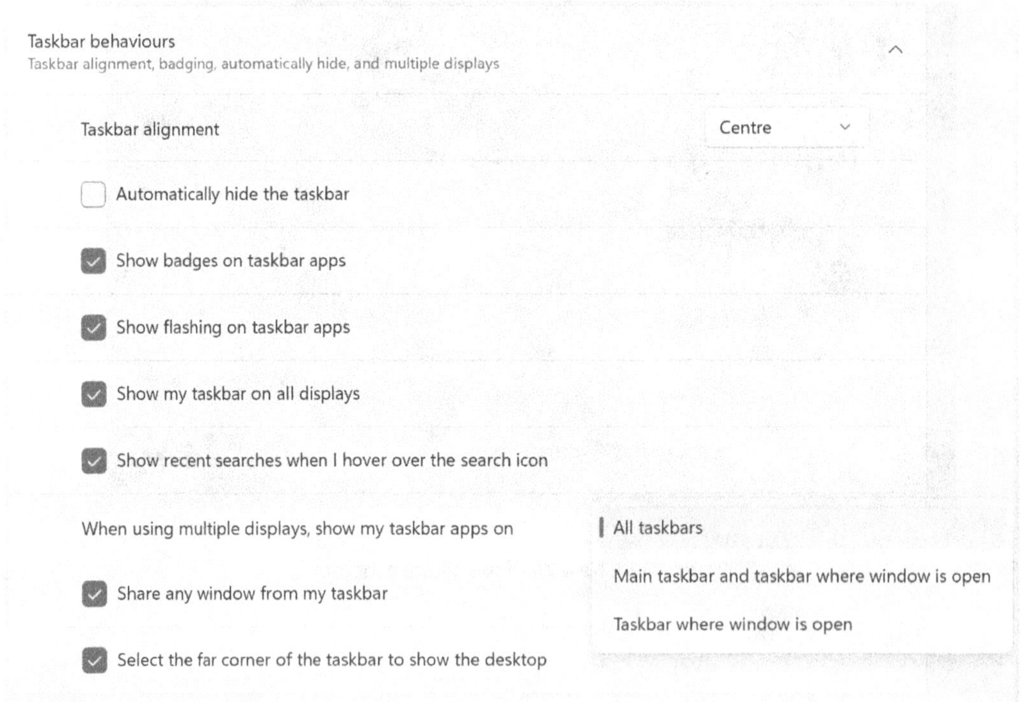

Figure 10.2 – New Windows 11 taskbar options

Some of the criticisms/missing features have been addressed with Windows 11 22H2; the number of pinned apps can now be changed and **Task Manager** is again accessible via a right-click. More about this later in the *Windows 11 22H2 (build 22621)* section.

New docking and multi-monitor experiences

Windows 11 has now more snap layout options to snap your app windows, and the control is easier to use. When you hover the mouse over a window's **Maximize** button, the new UI becomes visible. Depending on your screen resolution, screen size, and DPI setting, you will get different options. To use the new snapping options, you need a minimum screen resolution of 1920 effective pixels. On smaller displays, you get the option to split in half or quarter:

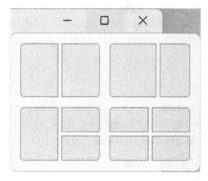

Figure 10.3 – Snap options on smaller displays

On larger screens, you also get the option of dividing the screen into equal thirds or thirds with a slightly larger center section:

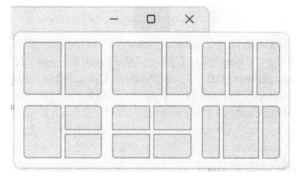

Figure 10.4 – Snap options on larger displays

> **Note**
>
> For the snap function to be displayed, the associated app must also support this feature. If you do not see this feature in some apps, it is not due to the graphics driver or OS settings. Even at the beginning of 2023, there are still apps, even from Microsoft, that do not currently support the snap feature. However, at least most Microsoft apps have support on the roadmap or are planned for the next major version of the program.

Besides the improved snap layout, Windows 11 has received a very pleasant function to remember monitor positions and the apps displayed on it. If you unplug an external monitor, the programs there are moved back to the main display. As soon as you plug the external monitor(s) back in, the apps are moved back to the external display. This way, you can comfortably take your portable device out of the docking station to participate in a meeting, for example, and when you return to your desk, you can usually continue working seamlessly without having to reposition the windows.

New patch file format

Windows 11 has a redesigned CU patching approach that results in up to 40% smaller size. At first glance, the CU patches are still delivered as MSU files, but they are just smaller in size now. However, there is a lot of improvement in the new internal structure of the MSU file. A very detailed article on how this size reduction could be done by removing reverse differential information can be read at `https://packt.link/hV7QC`.

In addition to the pleasant change that the CUs are now significantly smaller, the new structure results in some changes in handling. The **Servicing Stack Update (SSU)** is now directly integrated with the MSU, which simplifies the installation. This leads, however, to the fact that you can still install MSU files with `WUSA.exe` as before, but no longer uninstall them. For `DISM.exe`, there are no changes when working with MSU files; installing and uninstalling are still possible.

The situation is different if you want/need to handle **Cabinet (CAB)** files. The SSU is (currently) included as a standalone CAB, but the CU consists of CAB and the **Patch Storage File (PSF)**. **Deployment Image Servicing and Management (DISM)** and **Windows Update Standalone Installer (WUSA)** cannot handle CAB + PSF directly, but only if they are included in the MSU. There are external third-party tools on GitHub that can create an MSU from PSF and CAB, but it is strongly recommended to use the MSUs because the internal pack format could change at any time. According to information released to the Microsoft tech community, "*the format may evolve as we continue to innovate on improving cumulative update fundamentals.*" A further size reduction was achieved in 22H2; more on that in the next section.

"New" security options

Many of the "new" default security features were already optionally available under Windows 10.

> **Note**
>
> We are using quotation marks for *new* because these options are not really new. They were all present in Windows 10, but as an optional feature to activate. So, a lot of customers did not care about them and are now *surprised* by these "new" and default-enabled security options. It is intentionally "new" in the Windows 11 21H2 section and new without quotation marks in the Windows 11 22H2 section because in 22H2 new things were really introduced (and not only switched on/enforced).

Administrators who have already activated all possible protection features under Windows 10 will therefore have an easy upgrade to Windows 11. If you have been putting off topics such as **Credential Guard** (**CG**) and **Hypervisor-Protected Code Integrity** (**HVCI**) and are using older drivers and apps, this could now backfire.

HVCI is only automatically activated in Windows 11 21H2 for the very latest hardware generations (for example, Intel 11th Gen and newer) during new installations since it is assumed that newer compatible drivers are available here. With Windows 11 22H2, these changes and HVCI are activated with new installations on all supported CPU generations.

> **Note**
>
> To keep the compatibility as high as possible during the in-place upgrade, it has been decided *NOT* to automatically enable features such as CG and HVCI during this process. It is strongly recommended to check the compatibility of all apps and drivers; if they are compatible, then enable CG, HVCI, and so on explicitly via a **Group Policy Object** (**GPO**)/**mobile device management** (**MDM**) to have the best possible protection on Windows 11 and maintain identical behavior between in-place upgrade and new installation.

It is not recommended to switch off the new default security under Windows 11 to reduce it to Windows 10 default. This should only be a temporary and short-term exception until all programs can cope with the new security. In the current cyber-threat situation, any security that is usable is valuable. Also, new Windows 11 security technologies should be evaluated and implemented as soon as possible, as it is likely that some of these optional security features will become the *new normal* in the future.

Other technical changes in 21H2

We won't be able to go into every single technical change of Windows 11 here, because that would completely exceed the length of this chapter. We will therefore limit ourselves to technical changes that are important for the administrator in our opinion.

New ISO packaging

Besides the well-known Windows 10 **Software Development Kit** (**SDK**), **Assessment and Deployment Kit** (**ADK**), and **Windows Driver Kit** (**WDK**), Windows 10 is available in two ISO formats that cater to different language needs, as well as one ISO format for the **Feature on Demand** (**FoD**) ISO. Sometimes, it was not easy to figure out the language ISO in which a particular language was included. Since Windows 10 2004 (20H1) evolved only via *enablement packages*, the base ISOs for language, ADK, and so on remained the same. So, even for the newest iteration, Windows 10 22H2, you still need the original Windows 10 20H1 versions of the ISOs.

With Windows 11 21H2, a completely new set of ISOs was released, and the language ISOs and FoD ISO have been replaced/merged with a single combined **Languages and Optional Features** (**LoF**) ISO. This LoF ISO includes all supported languages and all optional features of the corresponding Windows 11 version. Windows 11 21H2 and Windows 11 22H2 use different ISOs, so please select the correct version.

Network drivers as FoDs

In the recent Windows 10 versions, more and more components were transferred to optional FoD v2 packages and had thus also become removable. Starting with Windows 11 21H2 preinstalled (out of the box), Wi-Fi and Ethernet drivers are now FoD packages. They are still preinstalled as default, but can now be easily removed to reduce disk footprint. This is especially helpful in scenarios involving **virtual desktop infrastructure** (**VDI**) or when small installation images are required. If the drivers are needed again later, they can be added just as any other FoD at any time.

Deprecated and removed features in 21H2

The following features now have a deprecated status in Windows 11 21H2 and will no longer be developed or improved. If you use one of these features, you should try to move away soon as they can be removed with one of the next Windows versions:

- **Cortana in Out of Box Experience** (**OOBE**): Cortana will no longer be used during the first boot experience. Also, it is no longer pinned to the taskbar. Cortana is still available as a standalone app.
- **BitLocker To Go Reader**: This feature was mainly used for older Microsoft OSs to access BitLocker-encrypted volumes. As all modern Microsoft OSs now support BitLocker out of the box, it is no longer needed and should not be missed.

- **Personalization roaming**: There will be no further development of roaming personalization settings, such as wallpaper, slideshow, accent colors, and lock screen images, across devices.

- **Windows Management Instrumentation Command Line (WMIC)** tool

- NTVDM/MS-DOS/16-bit support

The following features were already deprecated earlier and have now been removed from Windows 11 21H2:

- **Live Tiles** were removed and there are no more dynamic 1x1 to 4x4 tiles

- Start menu pinned locked apps/groups

- The **News and Interests** app has been removed and evolved into **Widgets**

- The Windows **Timeline** feature was removed from Windows 11 and partially replaced by timeline features in Edge

- XDDM-based remote display driver

- Microsoft Edge (legacy version)

- Internet Explorer 11 desktop application

- Math input panel

- Windows S mode (only available as Windows Home S)

For a regularly updated list with more descriptions, look at `https://packt.link/LTx6g`.

Windows 11 22H2 (build 22621)

The second iteration of Windows 11 was released in GA in September 2022 with build 22621.521 and officially branded as Windows 11 22H2 (or Windows 11 2022). Due to the release of the second version, the original Windows 11 was rebranded to Windows 11 21H2. Some weeks after the release of Windows 11 22H2, a new version of Windows 10 called 22H2 was also released. So, it is best to also mention the build numbers as they are in some cases the only differentiator programmatically for the different Windows 10 and 11 versions. Some major criticisms of the original Windows 11 version have been addressed with 22H2, which significantly improves the UI and user experience. These are discussed in the following sections.

New security options

With Windows 11 22H2, new security features and enhancements were introduced. We picked some important ones that you should be aware of, so let's have a closer look.

HVCI and VBS enabled by default

Windows 11 21H2 was already activating HVCI and **virtualization-based security** (**VBS**) for newer CPU generations on new installations by default. Now, with Windows 11 22H2, the OS activates HVCI and VBS on *ALL* supported CPU generations during the new install. There is no change of HVCI status for in-place upgrades where HVCI is not forcefully activated. Configure this feature per GPO/MDM to have the same behavior on in-place upgrades and new installations. (We would still recommend that this be left activated; see the *New security options section under Windows 11 21H2 (build 22000)*.)

Devices running Windows 11 Enterprise will also automatically activate Windows Defender Credential Guard running on VBS "*to greatly increase protection from vulnerabilities in the operating system and prevent the use of malicious exploits that attempt to defeat protections*," as conveyed to the Microsoft tech community.

Local Security Authority (**LSA**) protections are also enabled by default under certain conditions: `https://packt.link/dgYMj`.

Phishing protection

Beginning with version 22H2, Windows 11 and enhanced phishing protection in Microsoft Defender SmartScreen are helping you keep passwords safer. How? By automatically detecting when a user types a password into any app or website, determining in real time if that app or site has a secure connection to a trusted site, and warning the user at the moment if they need to change their password to reduce potential compromise to organizational resources.

Enterprise password

This enhanced phishing protection also automatically reports unsafe password usage to IT admins through the Microsoft Defender for Endpoint portal so that the incident can be tracked. Enhanced phishing protection also identifies and protects against password reuse on any app or site and typing or storing passwords in Notepad, WordPad, or Microsoft 365 apps. For a closer look at how this feature works—and how to configure it—see `https://packt.link/y3O1v`.

Personal Data Encryption

Windows 11 22H2 introduced a new security feature called **Personal Data Encryption** (**PDE**). Unlike BitLocker, PDE encrypts individual files rather than entire disks. So, PDE can and should be used in addition to other encryption methods such as BitLocker.

PDE uses Windows Hello for Business and releases encryption keys only when the user logs in with this security feature. PDE optionally discards encryption keys when the device is locked. Windows Hello for Business is used by PDE to secure the keys that protect the files. Previous **Encrypting File System** (**EFS**) technology uses certificates to secure the files, while PDE uses AES-CBC with a 256-bit key.

The current implementation of PDE has some requirements and limitations. You will need an Azure AD joined device (hybrid Azure AD joined not yet supported) with Windows Hello for Business and Windows 11 Enterprise or Education edition. It is not yet compatible with **Fast Identity Online** (**FIDO**)/security key authentication, Winlogon **automatic restart sign-on** (**ARSO**), **Windows Information Protection** (**WIP**), and Remote Desktop connections. Also, there is currently no GUI for PDE activation/control. To activate PDE, you need to configure an MDM policy, for which we can choose between two different security levels.

More details about PDE and the recommended additional security hardenings can be read here: `https://packt.link/slWi5`.

Even though PDE still has many limitations, it sounds like a promising additional transparent security layer and should definitely be kept in mind.

Smart App Control

Smart App Control (**SAC**) is a new security option in Windows 11 22H2 that adds additional security protection from new and emerging threats by blocking apps that are malicious or untrusted. SAC checks with an intelligent cloud-powered security service to make confidence predictions about the security of an app before executing it. If the service believes the app to be safe it will let it run; otherwise, it will be blocked before being able to harm the system.

SAC can only be activated on freshly installed Windows 11 22H2 devices. If you in-place upgraded to Windows 11 22H2, it is not possible to activate SAC without resetting the device. Once turned off manually by any user of the system, it is also not possible to reactivate SAC without this reset process. SAC will automatically turn off for enterprise-managed devices unless the user has turned it on first.

After a fresh clean install of Windows 11 22H2, SAC is in evaluation mode and will check if all prerequisites for running SAC are met and check for legitimate tasks that interfere with SAC too often. If such tasks are identified, SAC is automatically switched off. You can manually activate SAC without evaluation mode, but it is not recommended. SAC needs optional diagnostic data; if you turn off optional diagnostic data, SAC will also be deactivated.

More information about SAC can be found at `https://packt.link/ATUTg`.

Improved Start menu and taskbar

The new Windows 11 **Start** menu was often criticized for having 6x3 pinned app items that were scrollable but not changeable in number.

With Windows 11 22H2, it is now possible to display 6x4 app pins with the new **More pins** option or 6x2 app pins with the **More recommendations** option. Accordingly, the recommendations are reduced to two or expanded to eight. By default, it remains at 6x3 app pins and six recommendations:

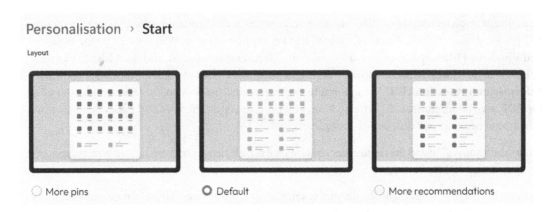

Figure 10.5 – New option to get more app pins or more recommendations in the Start menu

To further improve the pinned app section, there is a new option to create folders and group apps in the pinned area. As you are used to on your iPhone or Android phone, just drag and drop an app icon to another to create a new group or add it to an existing group. After creating a new group, you can name it by clicking on it once:

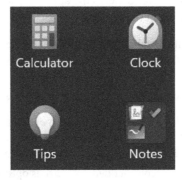

Figure 10.6 – Sample of a group with three apps

When you click on such a group, it automatically expands, and you can reposition the order of apps, drag apps out of the group, or rename the group:

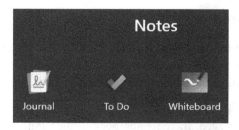

Figure 10.7 – Expanded "Notes" sample group

When moving out of the second-to-last app, the group will be automatically deleted.

Also, some missed drag and drop features are now back in the new taskbar, so it is possible to drag and drop without using the *Alt + Tab* workaround.

Additionally, the missing feature to access **Task Manager** with a right-click on the taskbar area was reintroduced in 22H2:

Figure 10.8 – New right-click option for Task Manager

Besides all the new features and changes to the security, the UI, and so on, we want to devote the next section to a topic that is personally very close to my heart—accessibility, and how Windows 11 provides significantly more integration here.

New inclusive features

Windows 11 has been redesigned in many areas. One of the areas with many improvements that we would like to highlight is the improved inclusive design. This has been optimized across the OS, as well as in the associated product suites and hardware to create an efficient, fun, and delightful experience for all people, whether with or without disabilities.

Contrast themes have been reimaged. There are now not only the familiar high-contrast themes from Windows 10 but also a large number of different contrast themes to adapt to different visual impairments. From the beginning, people with visual impairments were involved in the development process.

Also, the Windows sounds have been reworked to make them calmer and more informative. The startup sound has been reintroduced to signal that the PC is ready. The Windows sounds have been made less aggressive, and care has been taken to use the full spectrum of 250-8000 Hz to be more audible to all.

The **Accessibility** features were rebranded from **Ease of Access** within the **Settings** page. The **Accessibility** settings are now more easily available in the **Quick Settings** area on the bottom right:

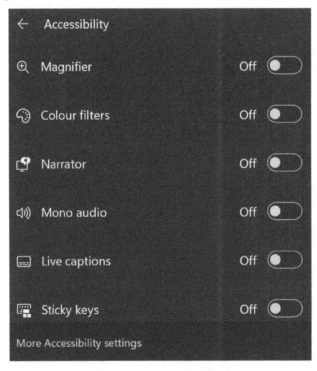

Figure 10.9 – Accessibility quick settings menu

For more information, see https://packt.link/YV9sc and https://packt.link/ycBfP.

Let's have a closer look at two new features—**Focus Assist** and **Live Captions on Windows 11 22H2**.

Focus Assist

The first powerful new feature is the **Focus Assist** feature, which is also often used as a tool to increase productivity since you are less distracted.

The new **Focus** assistant is easily accessible via the time/calendar function in the bottom-right corner. However, some of its settings and capabilities are well hidden. This tool ensures that you are not distracted during the concentration phase, but app notifications are not lost. This feature can be used by everyone:

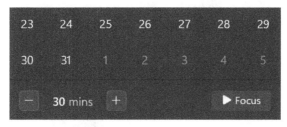

Figure 10.10 – Focus Assist quick start below the calendar

A small part of the settings can be found in **Settings | System | Focus**. There, you can define which notifications should be displayed or not, specify a session length, and also start it:

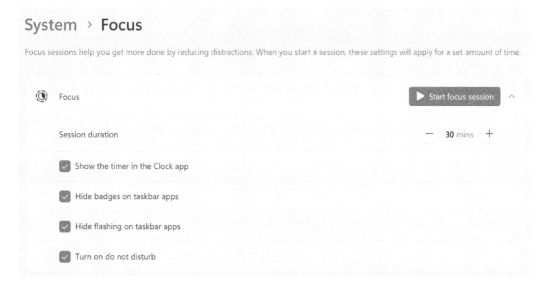

Figure 10.11 – Focus Assist settings on the Settings page

The really useful settings—for example, to schedule daily focus times, settings for the timing, and number of breaks, as well as an integration with **Microsoft To Do** and **Spotify** (for the favorite playlist during concentrated work)—are hidden in the **Clock** app under **Focus sessions**:

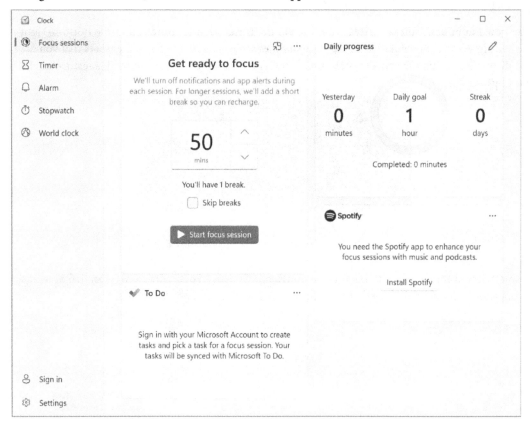

Figure 10.12 – Enhanced Focus Assist settings in the Clock app

This makes **Focus Assist**, which was originally aimed at people with ADHD, a productive tool usable by everyone.

Transcription/Live captions

The second powerful feature we would like to point out is **Live captions**. The tool has some advantages that may not be obvious at first glance. It can create live captions very well and independently of apps. It is freely configurable how and where the font is displayed. **Live captions** supports any audio content, no matter if it is a replayed video or a video conference. And here, we come to a feature that many may not be aware of. In the current pandemic times, it has become common practice to wear masks when meeting with many people in a conference room. However, this is a big problem for people who rely on lip reading. This is where it can help because you can also set **Live captions** to capture the microphone as an audio source.

This feature is not only helpful for people with hearing deficits but also for non-native speakers. If you are not a native speaker of the other person's language, it is helpful to have live subtitles, which can significantly improve comprehension.

Live captions is fast and easy to access via the **Accessibility** quick settings in the lower-right corner. The numerous settings and the activation of the microphone can be found on the **Settings** page under **Settings | Accessibility | Captions**:

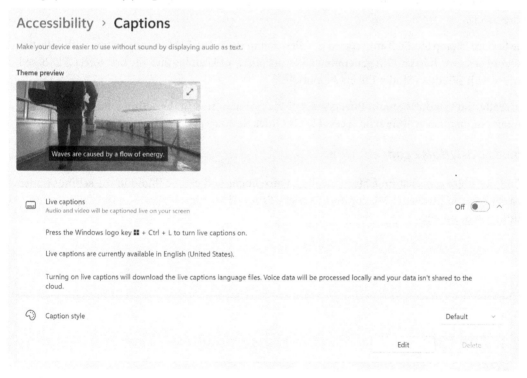

Figure 10.13 – Live captions settings page

When using **Live captions** for the first time, a small language-dependent file has to be downloaded.

Windows Studio effects

Windows Studio effects are multiple video and audio enhancements introduced with Windows 11 22H2. Currently, Windows Studio Effects are only available with special AI engines integrated into the CPU. They use highly specialized **compute engines**, also known as **neural processing units** (**NPUs**), to perform these complex calculations without loading the main CPU. This leads to significantly less CPU load even when using the effects, cooler systems during video conferences, and longer battery runtimes.

Windows Studio Effects are applied directly to the camera and/or microphone, so no special support from the app is needed since everything is already built into the OS.

The first product in the market was Qualcomm's Snapdragon 8cx Gen 3 with NPU, which debuted as the first system supporting Windows Studio Effects with Microsoft SQ3 in Microsoft's Surface Pro 9 5G. AMD announced at *CES 2023* that it would integrate its implementation of an AI engine (called Ryzen AI) in its new Ryzen "Phoenix" processors aka Ryzen Mobile 7040 series launched in Spring 2023.

Intel has announced that it will first integrate AI support in its next processor generation, named Meteor Lake. At Intel, these NPUs are called Intel **Versatile Processing Units** (**VPUs**). Microsoft has already announced that it will make Windows Studio Effects available on more devices as well as add more effects.

The Surface Laptop Studio 2 announced by Microsoft in the fall of 2023 in cooperation with Intel has a special position. It has a 13th generation CPU, but also a VPU and is thus the first Intel CPU-based Surface with Windows Studio Effects support.

Currently, the Windows Studio Effects are still somewhat buried in the settings, but Microsoft has announced that they will soon be accessible via **Quick Settings** at the bottom right of the screen.

Voice Focus/Noise cancellation

Windows' noise cancellation feature is called **Voice Focus** and can be found in the settings under **System | Sound | Internal Microphone Array - Front** (this point depends on the device name) | **Audio enhancements**.

Voice Focus eliminates background noise from the audio stream in real time. To use this tool, **Audio Enhancements** must be set to **Microsoft Windows Studio Voice Focus** (which should be automatically on supported devices). After that, **Voice Focus** can be set to **On**:

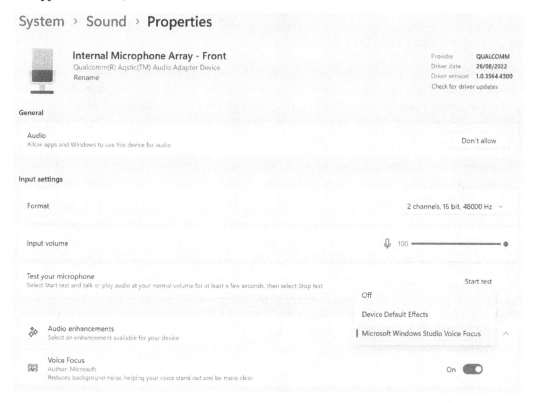

Figure 10.14 – Voice Focus settings/Audio enhancements on the Properties page of Microphone

To get a better idea of the performance of this feature, I recommend watching the funny video with the hair dryer and Steven Bathiche and the Surface Pro 9 5G as a demo (`https://packt.link/YAumM`) or the more serious version of the *Ignite* video about the new AI features with Panos Panay and Steven Bathiche supported by the guitarist Bob Bejan (`https://packt.link/eXmrT`).

New and improved video effects

Besides the audio effects, Windows Studio Effects also offer several new and improved video effects. **Automatic framing** and **Eye contact** were not available until now; background blur was heavily improved; and now, it also offers a portrait option. These effects are also well buried in the settings. They are located under **Settings | Bluetooth & devices | Cameras | Surface Camera Front** (this point depends on the device name) | **Camera Effects**.

The first feature is **Automatic framing**, which zooms in on the person in the picture. It is designed to always keep you centered in the video frame, even if you are moving around or the webcam is placed at an angle.

The second video feature is **Background effects**. You can choose between the classic **Standard blur**, where the background is heavily blurred and there is a harder transition between the person and the background, and the new **Portrait blur** with a smoother transition, less blurred background, and a more portrait look. Both **Standard blur** and **Portrait blur** work much cleaner and are more error-free than the CPU-based variant that was available until now. Finally, the third video feature is **Eye contact**. With the help of AI, a video image is calculated in which you always maintain eye contact with the other person (even if you read something and avert your gaze):

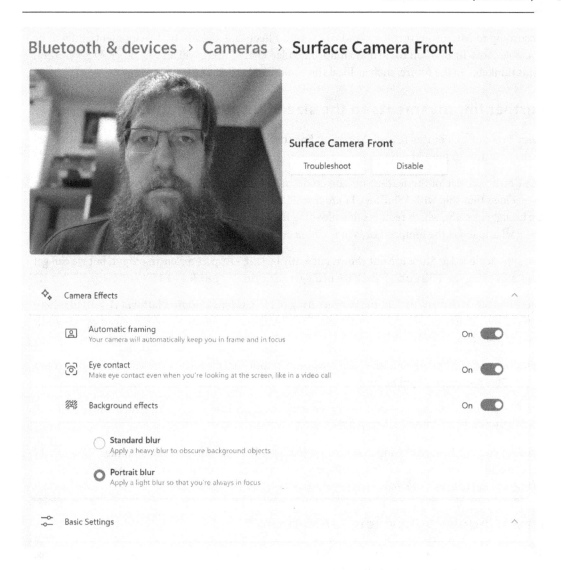

Figure 10.15 – Camera Effects on the Properties page of Camera

According to Microsoft, further Windows Studio Effects are already in development and will be available soon. In addition to further audio and video effects, there will also be other AI application fields thinkable in the future, such as local translation and much more.

Further improvements to the size and speed of updates

Every byte counts. For this reason, Microsoft has continued to work on the size of the image as well as the CUs in Windows 11 22H2.

The download size of the feature update could be significantly reduced by redesigning how the *in-box apps* that ship with Windows 11 are handled. Starting with Windows 11, a dozen inbox apps are brought as *stubs*, which requires downloading the matching full app when the app is first started. The goal is to lower the footprint of Windows images, which was reduced by ~1 GB with this change.

The apps deployed as Stubs are not shown currently by `Get-AppxPackage` output, but we can get the status using the `IsStub()` package manager (`https://packt.link/DAoWq`).

These and other changes such as the restructuring of the **Unified Update Platform** (**UUP**) files have reduced overall download size by ~450 MB. The install disk requirement was reduced by ~399 MB and the peak disk usage by ~849 MB.

But it was not only redesigned for space consumption but also for faster installation. This will result in ~21-30% reduction of offline time.

The CUs were also further streamlined by reducing the download size, making them faster to install while consuming less disk space and reducing CPU time to install.

Some of the details were already described in the *New patch file format* section under *Windows 11 21H2 (build 22000)*. For more facts, details, and backgrounds, have a look at *Faster. Smaller. Windows 11, version 22H2 update fundamentals*: `https://packt.link/rKPlK`.

New Windows Subsystem for Android™

The **Windows Subsystem for Android** (**WSA**) app enables Windows 11 to run Android applications. Microsoft cooperates with Amazon and utilizes the Amazon Appstore.

WSA was already introduced with Windows 11 21H2 22000, but for a long time, it was only available in Insider and only on `en-US`. With Windows 11 22H2, WSA was significantly improved and became GA in several regions, which is why we now list it in the *Windows 11 22H2 (build 22621)* section. Currently, WSA is still based on Android 12; a version based on Android 13 is on the Insider channel right now.

To officially install WSA, you have to go to the **Microsoft Store**, find the **Amazon Appstore** app, and install it:

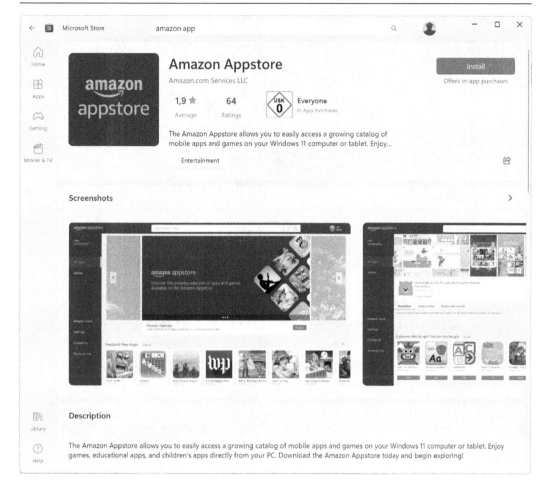

Figure 10.16 – Amazon Appstore in the Microsoft Store

> **Note**
>
> There is also the possibility to install only WSA via UUP and the command line and install other app stores via debug. Since this is not officially supported by Microsoft, I will leave it as a mention and not show any details.

The installation of the Amazon Appstore is a multi-step installation, depending on whether you don't yet have Hyper-V features installed (three steps) or whether Hyper-V is already set up (two steps).

In the three-step scenario, it will first try to install the Hyper-V feature:

Figure 10.17 – Step 1 to activate virtualization

To do this, you need to have admin rights. By this, the activation of WSA is automatically blocked for all non-admin users (there is no extra GPO or MDM to block it, but if your users don't have admin rights, it is all blocked). Activation of Hyper-V needs a reboot, and the process continues right after it.

In *step 2* (which would be *step 1* in the two-step scenario), WSA itself is being installed.

In *step 3*, the Amazon Appstore app is installed into WSA.

To control WSA, there is an extra **Windows Subsystem for Android settings** app, where you can set the startup behavior, control graphics and performance, reset to defaults, and turn off WSA.

After successfully installing WSA together with the Amazon Appstore, you need to log in to the Amazon Appstore. For this, a private Amazon account is needed. Microsoft accounts and Azure AD accounts are not supported.

At the time of writing this book, the Amazon Appstore offered more than 95% of games and only some apps such as media libraries. The lack of WSA controls via GPO/MDM, the need for a private Amazon account, and the current app offer make it more of a consumer rather than a business feature. Perhaps—or better, hopefully, this will change in future developments. Therefore, we will not describe WSA in full detail here. If you are interested in WSA details, please have a look at `https://packt.link/UJVtB`.

CI aka Moments

Windows 11 22H2 introduced a new approach from Microsoft to deliver features. This approach was called **Moments** and is now called **CI**. Besides the well-established delivery with now yearly **feature updates**, it offers the possibility to add features through periodic updates (aka CUs) and Microsoft Store updates for apps. You will have noticed the original GA version of Windows 11 22H2 missed some features announced earlier in 2022 such as the **Explorer** tabs. These tabs were introduced with the so-called October Moment, the optional October CU, and all following CUs.

Opinions on this change in how features will be included in Windows 11 could not be more different. Many welcome the fact that features that were not ready/final on the GA's completion date no longer have to wait a whole year and can be delivered later when they are ready. However, there is also a clear critique from the enterprise environment that the features are delivered in the CU, must be introduced within a short time, and cannot be explicitly switched on/off. It is also unclear what is planned for the course of the year.

The contents of the October Moment have already been documented together with the GA release note. It has also already been announced that there will be more such updates. However, Microsoft has not yet committed to a definitive number of Moments as of January 2023, nor has it documented which features will be included in the next Moment.

Microsoft addressed the criticism that new features introduced by these Moments were enabled by default and there were no options to do a controlled rollout. They stated that they will "*ship features off by default and create a single policy (GP/MDM) that allows enterprise customers to turn on these features. This will be done as a set, and not for individual features or individual releases.*" You can read more about this here: `https://packt.link/4h4sT`.

However, this control will only come for future features introduced after November 2022. With the introduction of Moment 2 in March 2023, the controlling GPO and MDM CSP were introduced to the OS and officially documented. All new features are off by default if the client gets it updated via **Windows Server Update Services (WSUS)** or **Windows Update for Business (WUfB)**.

Other technical changes in 22H2

As we already mentioned in the *Windows 11 21H2* section, we aren't able to go into every single technical change in this chapter, and therefore limit ourselves to the category of *other technical changes* that includes those that are important for the administrator/user.

Removal of WinPE x86

Since Windows 11 no longer supports x86-only OS installations, the next logical step was to also remove support for WinPE x86 in the Windows 11 22H2 ADK. Most deployment tools can cope with this/ have been adapted to it, with one prominent exception. Unfortunately, the **Microsoft Deployment Toolkit (MDT)** has not been adapted to this yet, and therefore older ADKs have to be used until further notice. Whether MDT will receive official support for Windows 11 in general or any future versions is currently unclear.

Availability of Explorer tabs

In the original GA version 22621.521, **Explorer** tabs were not included and were added with the October Update Preview CU 22621.755 as October Moment. All Windows 11 22H2 versions with patch level .755 or higher should have **Explorer** tabs enabled automatically. As known from Edge and other products, you can now open multiple tabs side by side and conveniently switch between them:

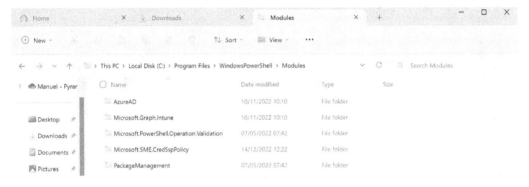

Figure 10.18 – The new tab view in Explorer.exe

As you can already see in the **Windows Insider Program**, such tab views are also planned for other built-in apps.

New UI for Task Manager

Task Manager has also been given a revised UI with a more modern look. The most important change is the menu on the left side. The settings can now be found at the very bottom left, represented by the cogwheel:

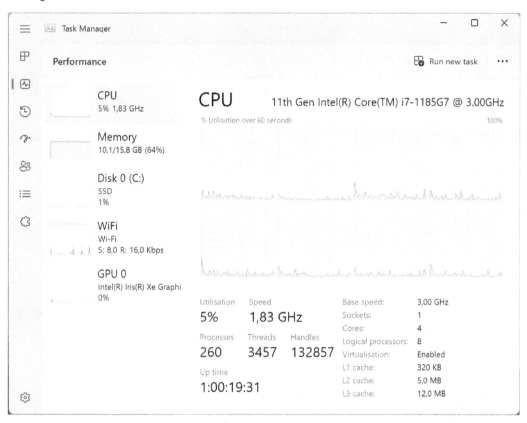

Figure 10.19 – New Task Manager UI

Here is a more detailed look at the expanded menu:

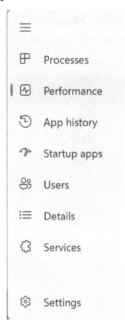

Figure 10.20 – Expanded Task Manager menu

There is only one new function for the time being; the other changes are currently limited to the visual appearance. This new function is **Efficiency mode**.

Improvements to Efficiency mode

The **Efficiency mode** feature itself was already introduced a few Windows 10 versions ago. Since then, it has been steadily improved, and apps such as Edge Chromium have also received a built-in variant of **Efficiency mode** (you will find this setting under the name **Put tabs to sleep automatically**). Gamers were also given the option to automatically improve their gaming experience with this efficiency control.

However, the possibility of manually sending a program into **Efficiency mode** was missing until now. Now, for each individual process, you can either right-click on the process and select **Efficiency mode** or you can left-click on the process and select the **Efficiency mode** icon at the top to start this mode:

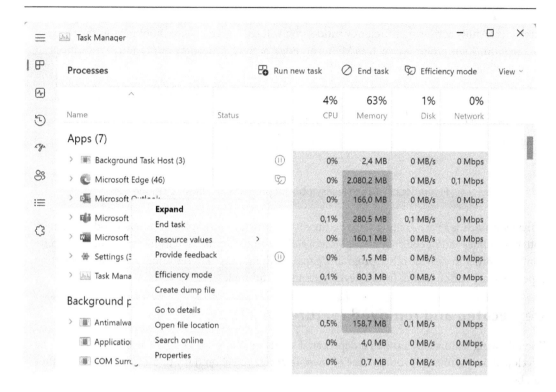

Figure 10.21 – Efficiency mode in Task Manager

If you want to switch a program to **Efficiency mode**, you will be shown a message with the consequences (lower priority, possible instability), which you have to confirm first:

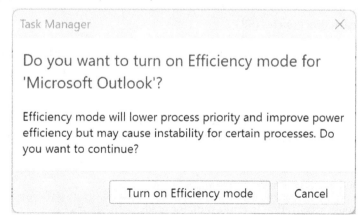

Figure 10.22 – Efficiency mode confirmation dialog

After successfully activating **Efficiency mode**, you will see a green leaf symbol next to the name of the program. Some programs, such as Microsoft Edge, activate **Efficiency mode** automatically or can be set to a higher efficiency in the settings:

Figure 10.23 – Green leaf symbol for programs in Efficiency mode

Efficiency mode is a very exciting topic because it can provide more system efficiency and thus longer battery runtimes and/or cooler systems. Moreover, Microsoft's approach to using this mode for priority control in order to provide other programs/games with more system resources is also an interesting aspect. We are curious about future developments on this topic.

Deprecated and removed features in 22H2

The following features are now in a deprecated status in Windows 11 22H2 and will no longer be developed or improved. If you use one of these features, you should try to move away soon as they can be removed with one of the next Windows versions:

- WIP—replaced by **Microsoft Purview Information Protection**
- **Windows Update Compliance**—replaced by **WUfB reports**

The following feature was already deprecated earlier and is now removed from Windows 11 22H2:

- **Software Restriction Policies (SRP)**

For a regularly updated list with more descriptions, look at `https://packt.link/qH4Rg`.

Other changes

Besides these new features, there are a lot more minor and major improvements and new features in Windows 11 to explore. To get an impression of what else has changed, here is a comprehensive list of additional things:

- Optimization for games in windowed mode
- Improved mouse acceleration in multi-monitor mode
- Improved **Airplane mode** remembers the status of Wi-Fi and Bluetooth
- Improved printer and scanner pages
- Discovery of Designated Resolvers support for encrypted DNS

- Improved VPN GUI to show more details

- Extended sharing improved with new UI

- Power and battery with recommendations to lower CO_2 footprint

- Improved storage interface

- Improved font installation UI

- Improved installed apps UI

- **Server Message Block (SMB)** compression

- Better **Dynamic Refresh Rate (DRR)** for screens on laptops with 120Hz

- Better integration of OneDrive in Explorer and Settings

- Suggested actions (currently en-US only)

- Personalized emojis

- Clipchamp video editing app

- Voice access (currently en-US only)

- Improved automatic DND mode

And there are many more.

Install the new Windows 11 22H2 as soon as possible and familiarize yourself with all the cool new features!

Further references for Windows 11 22H2

There are a lot of helpful and interesting-to-read links and blogs around Windows 11 22H2, which we try to summarize here:

Visit *What's new for IT pros in Windows 11, version 22H2* (https://packt.link/nDmXN) for additional information about 22H2, including the following:

- Deploying Windows 11, version 22H2

- Updated IT toolbox

- Windows ADK for Windows 11, version 22H2

- New and updated Windows release information experiences

- Moving from Windows 10 to Windows 11, version 22H2

- Monitoring the status of your rollout

- Servicing life cycle

Here are some additional links:

- *Available today: the Windows 11 2022 Update* (`https://packt.link/dW1Ea`)

- *How to get the Windows 11 2022 Update* (`https://packt.link/PoUUi`)

- *Work safer and smarter with Windows 11 2022 Update* (`https://packt.link/FaqXw`)

- *New Windows 11 security features are designed for hybrid work* (`https://packt.link/DexSp`)

- *Protect passwords with enhanced phishing protection* (`https://packt.link/hM0Lb`)

- *Faster. Smaller. Windows 11, version 22H2 update fundamentals* (`https://packt.link/p4ubj`)

- *How inclusion drives innovation in Windows 11* (`https://packt.link/zmjCb`)

- *Windows Update is now carbon aware* (`https://packt.link/dq1SU`)

- *Inside this update* (`https://packt.link/aY2Qy`)

Here are some links to official blogs on the subject:

- *What's new for IT pros in Windows 11, version 22H2*: `https://packt.link/jn26b`

- *Work safer and smarter with the Windows 11 2022 Update*: `https://packt.link/kpPmS`

Summary

In this chapter, you have learned about the changes in Windows 11 21H2 and 22H2. This should help you decide which version to use and which features to look at closely during piloting or to include in design decisions. We have not only introduced you to the obvious changes in the UI but also to the changes in the Windows substructure. We have summarized a selection of important changes, but as you can see from the numerous references to URLs and other changes, the changes are much more extensive. I hope we have aroused your interest in the new operating system. Take the opportunity to try out Windows 11 22H2 as soon as possible and get an impression of the new operating system.

Index

Packtpub.com

Subscribe to our online digital library for full access to over 7,000 books and videos, as well as industry leading tools to help you plan your personal development and advance your career. For more information, please visit our website.

Why subscribe?

- Spend less time learning and more time coding with practical eBooks and Videos from over 4,000 industry professionals

- Improve your learning with Skill Plans built especially for you

- Get a free eBook or video every month

- Fully searchable for easy access to vital information

- Copy and paste, print, and bookmark content

Did you know that Packt offers eBook versions of every book published, with PDF and ePub files available? You can upgrade to the eBook version at packtpub.com and as a print book customer, you are entitled to a discount on the eBook copy. Get in touch with us at customercare@packtpub.com for more details.

At www.packtpub.com, you can also read a collection of free technical articles, sign up for a range of free newsletters, and receive exclusive discounts and offers on Packt books and eBooks.

Other Books You May Enjoy

If you enjoyed this book, you may be interested in these other books by Packt:

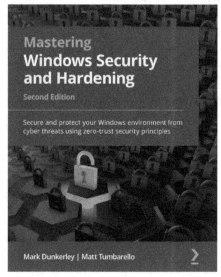

Mastering Windows Security and Hardening - Second Edition

Mark Dunkerley, Matt Tumbarello

ISBN: 978-1-80323-654-4

- Build a multi-layered security approach using zero-trust concepts
- Explore best practices to implement security baselines successfully
- Get to grips with virtualization and networking to harden your devices
- Discover the importance of identity and access management
- Explore Windows device administration and remote management
- Become an expert in hardening your Windows infrastructure
- Audit, assess, and test to ensure controls are successfully applied and enforced
- Monitor and report activities to stay on top of vulnerabilities

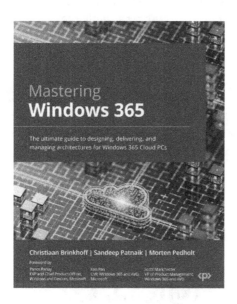

Mastering Windows 365

Christiaan Brinkhoff, Sandeep Patnaik

ISBN: 978-1-83763-796-6

- Understand the features and uses of Windows 365 and Cloud PCs
- Extend your existing skillset with Windows 365 and Intune
- Secure your Windows 365 Cloud PC connection efficiently
- Optimize the Cloud PC user experience through effective analysis and monitoring
- Explore how partners extend the value of Windows 365
- Use the available tools and data within Windows 365
- Troubleshoot Windows 365 with effective tips and tricks

Packt is searching for authors like you

If you're interested in becoming an author for Packt, please visit authors.packtpub.com and apply today. We have worked with thousands of developers and tech professionals, just like you, to help them share their insight with the global tech community. You can make a general application, apply for a specific hot topic that we are recruiting an author for, or submit your own idea.

Share Your Thoughts

Now you've finished *Windows 11 for Enterprise Administrators*, we'd love to hear your thoughts! Scan the QR code below to go straight to the Amazon review page for this book and share your feedback or leave a review on the site that you purchased it from.

https://packt.link/r/1804618594

Your review is important to us and the tech community and will help us make sure we're delivering excellent quality content.

Download a free PDF copy of this book

Thanks for purchasing this book!

Do you like to read on the go but are unable to carry your print books everywhere? Is your eBook purchase not compatible with the device of your choice?

Don't worry, now with every Packt book you get a DRM-free PDF version of that book at no cost.

Read anywhere, any place, on any device. Search, copy, and paste code from your favorite technical books directly into your application.

The perks don't stop there, you can get exclusive access to discounts, newsletters, and great free content in your inbox daily

Follow these simple steps to get the benefits:

1. Scan the QR code or visit the link below

https://packt.link/free-ebook/9781804618592

2. Submit your proof of purchase
3. That's it! We'll send your free PDF and other benefits to your email directly

www.ingramcontent.com/pod-product-compliance
Lightning Source LLC
Chambersburg PA
CBHW080631060326
40690CB00021B/4887